ANTOINE LAVOISIER

SCIENTIST • ECONOMIST • SOCIAL REFORMER

THE DA CAPO SERIES IN SCIENCE

Lavoisier and Madame Lavoisier from a portrait painted by David in 1788. *Bulloz, Paris.*

ANTOINE LAVOISIER

SCIENTIST • ECONOMIST • SOCIAL REFORMER

DOUGLAS McKIE

A DA CAPO PAPERBACK

Library of Congress Cataloging in Publication Data

McKie, Douglas.
Antoine Lavoisier: scientist, economist, social reformer / by Douglas McKie.
p. cm.—(Da Capo series in science) (A Da Capo paperback)
Reprint. Originally published: New York: H. Schuman, 1952.
Includes bibliographical references and index.
ISBN 0-306-80408-5
1. Lavoisier, Antoine Laurent, 1743–1794. 2. Chemists—France—Biography. I. Title. II. Series.
QD22.L4M32 1990
540′.92—dc20 90-38768
[B] CIP

This Da Capo Press paperback edition of *Antoine Lavoisier* is an unabridged republication of the edition published in New York in 1952. It is reprinted by arrangement with Harper & Row, Publishers, Inc.

Published by Da Capo Press, Inc.
A Subsidiary of Plenum Publishing Corporation
233 Spring Street, New York, New York 10013

Manufactured in the United States of America

CONTENTS

ANTOINE LAVOISIER

SCIENTIST • ECONOMIST • SOCIAL REFORMER

I

INTRODUCTION

AS OUR MODERN CIVILIZATION ADVANCES FROM YEAR TO YEAR, almost from day to day, we are steadily becoming more aware that science is increasingly affecting and influencing our daily lives, the structure of our thought, and our understanding of the world around us. The speed of scientific progress has now grown rapid and its application almost immediate; but the pace was once more leisurely.

The change from ancient to modern in science began in the sixteenth, and became evident in the seventeenth century, in an altogether different social context, in which the professional scientist had not yet made his appearance; and the fruits of some three hundred years of experiment and research have in our own time come amazingly to harvest. More than fifty years ago, before this was so plainly to be seen, the British statesman and philosopher, Lord Balfour, remarked of scientists: "They are the people who are changing the world and they don't know it. Politicians are but the fly on the wheel—the men of science are the motive power." To say nothing of invention and discovery and to refer only to everyday matters—we see that in the preservation of health, in the fight against disease and epidemics, in diet and in problems of malnutrition, in the production and preservation of food, in the preparation of artificial materials and fabrics and synthetic drugs and agricultural fertilizers, in economizing old and tapping new sources of power—to name only a few: in all these science is now the guide of civilized mankind and the dominant factor in the preservation of the military security of nations.

Yet science is not the mere search for new inventions, not a matter of radio and television and motorcars and weapons and gadgets, but the pursuit of a knowledge of Nature for its own sake and for a better understanding of the world in which we live and of its mechanism and of ourselves as part of it. It is to the minds that were inspired with the vision of searching into every nook and cranny of Nature that we owe much of what we now enjoy and, by its misapplication, occasionally suffer.

If we take a backward glance, we discover three great revolutions in science, each of them momentous in the history of Civilization: the revolution in mechanics in the seventeenth century, completed by Newton and marked by the publication in 1687 of his *Philosophiæ Naturalis Principia Mathematica* or *Mathematical Principles of Natural Philosophy;* the revolution in chemistry in the eighteenth century, effected by Lavoisier and signalized by the appearance of his *Traité Élémentaire de Chimie* or *Elements of Chemistry* in 1789; and the revolution in biology in the nineteenth century, achieved by Darwin and introduced in his *Origin of Species* in 1859. The three books in which these fundamental advances were recorded have done more than all others to mold our modern world; in their time, they gave the shape of things to come.

Newton formulated the mathematical laws relating to the motions of all bodies, terrestrial and celestial—of the apple falling from the tree and of the planets revolving about the sun, the moon in her orbit round the earth and the comets coming from and returning to the far spaces of the heavens. Darwin produced the evidence for evolution, but could only theorize about its mechanism, which he ascribed to "natural selection" or the "survival of the fittest," and it was left to those who came after him to offer a more satisfactory explanation of the facts that he had incontrovertibly established.

Lavoisier changed the whole structure and outlook of chemistry, the science which more than any other touches and affects our daily lives. It is the purpose of this book to give some account of his life and work and personality, of his problems and of the times in which he lived, and the context of his varied achievements.

The chemist's elements, the primary substances that in various numbers and proportions compose the vast multitude of different materials in the world around us, now number over ninety; and the names of many of them have become everyday words—oxygen, nitrogen, radium, and, more recently, uranium. But science took many centuries to arrive at a knowledge of what substances are to be regarded as elements, as simple uncompounded bodies.

From the times of the Greeks until almost the end of the eighteenth century—in the long vista of history, until yesterday—all substances were believed to be composed of only four elements, earth and water, air and fire, combined in different proportions to produce the extraordinary variety of materials in Nature. This four-element theory was first propounded by Empedokles of Akragas in Sicily, who lived from 490 to 430 B.C., and, although it was criticized as unsatisfactory after the rise of modern science in the sixteenth and seventeenth centuries, it was generally accepted and even passed into the common belief and speech of everyday life. It still survives among us today—we speak of a storm as "the raging of the elements;" "Does not our life consist of the four elements?" asked Sir Toby Belch in *Twelfth Night;* and, more recently, Winston Churchill described a war commander as "equally at home in the three elements, earth, air and water—and he is also well accustomed to fire."

As for combustion, among the most ancient of human observations, it was supposed that when a substance burned, some "fire-stuff," the very element of fire, was being released from it in the form of fire and flame, and that the substance was thus decomposing into its elements. This explanation was reasonable in its time, to judge by appearances only; most of us today, lacking the knowledge that we possess through later discoveries, would be quite ready to agree that when a match is struck or a candle or a coal burns some "fire-stuff" is released from each of them, and similarly from all other burning bodies.

The four-element theory was accepted for well over two thousand years and even after Robert Hooke (1635-1703) had contended in his *Micrographia* (London, 1665) that fire was not an element, but a mixture of air and combustible matter. It was Lavoisier who showed in the closing decades of the eighteenth century that Nature was not constructed on so simple a basis as four elements, and that bodies burn, not by releasing their imagined content of "fire-stuff," but by combining with the element oxygen, which forms one-fifth of the common air that we breathe. By these labors and what followed from them he created modern chemistry.

Moreover, apart from contributing to the development of other sciences, such as physics, geology, and cartography, Lavoisier was actively concerned in the application of science to industry and manufactures and especially to agriculture; and, unlike Newton and Darwin, he was prominent as economist and social reformer in the affairs of his country. Much that he urged in the organization and application of science has been realized only in our own time; and much that he strove for in the troubled politics of eighteenth-century France might, had it come to pass, have saved his countrymen from the more desperate remedies to which they were subjected.

The narrative of Lavoisier's scientific achievements and of his studies in agriculture, economics and social reform, here presented is based throughout on his original publications, all of which are now rare, but which have been reprinted in the magnificent collected edition of his *Œuvres* in six large quarto volumes (Paris, 1862-93), published by the Government of France under the supervision of the Ministre de l'Instruction Publique in reparation of the memory of one of the most illustrious victims of the Revolution. Much of the biographical content is derived from the authoritative work of one of Lavoisier's fellow-countrymen, Édouard Grimaux (1835-1900), Member of the Académie des Sciences and Professor of Chemistry in the École Polytechnique. His book, *Lavoisier, 1743-1794* (Paris, 1888), is an account of Lavoisier's life and wider activities, with only one summarizing chapter on his scientific work, and embodies the results of Grimaux's researches in the large collection of documents and personal papers preserved after Lavoisier's death. For Grimaux's long and patient toil the author can only express his indebtedness and admiration after reviewing as much of the material as was readily accessible.

The author is pleased to record his gratitude to the Académie des Sciences, the Bibliothèque Nationale, the Archives de France, the Musée du Louvre, and the Musée Carnavalet for access to their collections and for permission to reproduce a number of illustrations from the originals in their possession.

II

EARLY YEARS

Birth and descent

ANTOINE LAURENT LAVOISIER WAS BORN IN PARIS ON MONDAY, August 26, 1743, in the Cul-de-sac Pecquet and christened, according to devout Catholic custom, on the day of his birth in the parish church of St.-Merry. He was baptized Antoine because the eldest sons of the Lavoisier family had, with one exception, borne this name for over two hundred years (which is as far as we know their history), and Laurent, after his godfather and great-uncle, Father Laurent Waroquier, Principal of the Catholic College at Beauvais. The Cul-de-sac Pecquet (Plate 1) is now a throughway, the Passage Pecouay, connecting the Rue des Blancs-Manteaux with the Rue de Rambuteau, near the Archives Nationales; and the house in which Lavoisier was born is no longer standing, but the narrow street preserves an air of older days.

Lavoisier's ancestors, who had risen in the social scale from humble origins, had lived and toiled for many generations in Villers-Cotterets, a small country town some fifty miles northeast of Paris, on the edge of the great forest of

1. Part of the Marais in the eighteenth century, showing the Cul-de-sac Pecquet, Lavoisier's birth-place, on the left off the Rue des Blancs-Manteaux, the Church of St.-Merry in the centre foreground, and the old Hôtel de Ville, overlooking the Place de Grève, on the right near the river. The house in which Lavoisier was born was probably the one at the end built across the axis of the cul-de-sac. *Musée du Louvre. From the 'Plan de Turgot,' made by order of Michel Étienne Turgot, prévôt des marchands, drawn by Louis Bretez from 1734–39 and engraved by Claude Lucas.*

Villers-Cotterets. The earliest of them of whom any record survives was Antoine Lavoisier, who died in 1620; he was a postillion in the King's service. His son, Antoine, was postmaster at Villers-Cotterets and died at some date before 1637; and his grandson, also Antoine, who was born about 1606 and died in 1691, became a sheriff's officer. Nicolas, son of the third Antoine, was born about 1641, became a merchant and prospered; and his son, Antoine, born in 1678, turned to the study of law, became attorney to the bailiwick of Villers-Cotterets and married Jeanne Waroquier, born in 1680, daughter of Jacques Waroquier, notary and attorney. The son of this marriage was Jean Antoine Lavoisier, also a lawyer, who in 1741 succeeded his uncle, Jacques Waroquier, and inherited his uncle's fortune, his place as attorney to the Parliament of Paris, and his house in the Cul-de-sac Pecquet.

On June 14, 1742, Jean Antoine Lavoisier married Émilie Punctis, daughter of Clément Punctis, advocate to the Parliament and secretary to the Marquis de Châteaurenault, vice-admiral of France. The first child of this marriage was Antoine Laurent Lavoisier, the founder of modern chemistry, born in those comfortable middle-class circumstances to which his family appear to have been accustomed from the time when his great-grandfather Nicolas turned successfully to trade and thereby enabled his son and grandson to enter the legal profession and to marry the daughters of wealthy lawyers in official positions. But both the first and last of the Lavoisiers bore the name of Antoine; throughout his life, the subject of this biography preserved his family connection with the little town of Villers-Cotterets, where his earliest known ancestor six generations before had ridden the horses of Henry IV and Louis XIII.

The first five years of Lavoisier's life were passed in the house in which he was born. The Marais, as the district was called, was a quiet part of Paris, and nearby were a monastery and several aristocratic town mansions with great gardens. There would be drives and walks to the river to see the boats with cargoes from Le Havre and Honfleur and Rouen, and to the palaces, the Tuileries and the Palais-Royal, to see the ceremonies and the guards. Then the only other child of his parents' marriage, Marie Marguerite Émilie, was born in 1745. Three years later, in 1748, Mme. Lavoisier died and Jean Antoine gave up his house and went to live with his two young children at the house of their grandmother, Mme. Punctis, (their grandfather Punctis having recently died), in the Rue du Four-St.-Eustache, now the Rue de Vauvilliers, between the Bourse du Commerce and Les Halles. The house cannot be identified; many of the houses in the street were demolished for the construction of Les Halles, the great market of Paris, but some remained and were still remaining a half-century ago. It was only a short distance from the Cul-de-sac Pecquet, about half a mile or so, and the children were still near the river and the great palaces. With Mme. Punctis lived her younger daughter, Marie Marguerite Constance, aged twenty-two in 1748, who became so devoted to the two motherless children that, it is said, she abandoned all thoughts of marriage in order to care for them.

For twenty-three years, until his marriage in 1771, this was to be Lavoisier's home. Jean Antoine Lavoisier, although still in his thirties, did not marry again; he seems to have turned to his son for companionship and as the boy grew to manhood there developed between father and son a remarkable friendship, which played a great part in molding the younger man's life and character. Indeed, it may well account for the fact that Lavoisier seems to have been a man of few friends, a man whose human relationships were satisfied by those in his own family circle and his home.

Education

UNFORTUNATELY, WE KNOW VERY LITTLE ABOUT LAVOISIER'S boyhood except for some references to visits to Villers-

Cotterets, vacations spent there and rambles in the forest. His father had been educated at the Collège des Quatre-Nations, also known as the Collège Mazarin (Plates 2 and 3) because it had been founded under the will of Cardinal Mazarin (1602-61), who had governed France during the minority of Louis XIV. It originally had places

2. The Collège Mazarin, seen across the river from the Jardin de l'Infante, the garden between the Louvre and the river. The building is now the Institut de France and the Academy of Sciences is one of the national academies housed in it. *Musée Carnavalet.*

for thirty young noblemen from the four provinces reunited to France by the treaties of Münster and the Pyrenees, and developed into a good school, probably the best in Paris. In the middle of the eighteenth century, however, the students were mostly day-boys from Paris, including, probably in 1754, the young Antoine Lavoisier. His way to school would carry him over the river by the famous Pont-Neuf (Plate 4), with the cathedral church of Paris, Notre-Dame, on his left, and the statue of "good King Henry" on his right—the statue to which the mobs of Paris at no very distant date would be compelling all

3. A court and garden in the Collège Mazarin. *Musée Carnavalet.*

aristocrats crossing the bridge to bow, and to descend from their carriages to render this courtesy to a well-remembered monarch whose ideal for every French peasant was "a fowl in his pot on Sunday." On his left, too, was the Conciergerie, the grim prison and anteroom for those awaiting trial at the Palais de Justice; but the shadow of these lay forty years ahead.

During Lavoisier's school years which probably ran

from 1754 to 1760, he gained a number of awards, mostly for Greek and Latin and French composition. At the end of what seems to have been his second year in the school, he was granted a special prize for industry in his studies, the first indication of that tireless energy that was later to impress not only his own generation, but all those who have

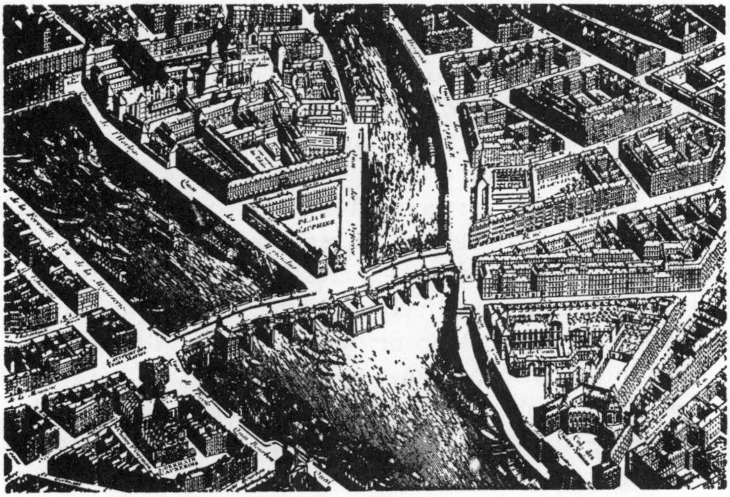

4. Another part of eighteenth-century Paris, showing the Collège Mazarin, or Collège des Quatre-Nations, in the right foreground near the river, and also the Pont-Neuf, the Conciergerie, and the Palais de Justice. *Musée du Louvre, Plan de Turgot.*

since become familiar with his work. His first interests at the Mazarin College were literary and, like many other men of science in their early years, he had dreams of becoming an author. He began a play, called "The New Héloïse," but completed only the first few scenes. He submitted essays for competitions arranged by the provincial academies, on such subjects as whether uprightness was as necessary as accuracy in the search for truth, and whether the desire of perpetuating one's name and deeds in human history was natural and reasonable—topics that he must have pondered over, perhaps not without some abiding influence on his character. In 1760 in the *concours général,* the annual competition for all the public schools of France, he was awarded the second place for rhetoric. Prizes such as these presented at the Sorbonne, the University of Paris, were soon to be gained by a younger contemporary, especially for his work in Latin; his name was Isidore Maximilien Robespierre, and one day their paths would cross.

The year 1760 was marked also by sorrow, for Marie Lavoisier died at the age of fifteen. In his home, however, Lavoisier's intelligence, his success at school, his enthusiasm for study, and his lively and affectionate ways helped to console the three older members of the family, who now seemed to live only for him.

The literary phase soon passed and young Lavoisier began to take an interest in science, to which he received his introduction at the Mazarin College in the courses of scientific lectures given to the senior classes during the year devoted to philosophy. The time, however, had come when he must choose a profession. Following family tradition and doubtless after discussion with his father, he entered the School of Law on leaving the Mazarin College, qualifying as a Bachelor of Law in 1763 and Licentiate in 1764. The courses at the School of Law were not so arduous as to leave him without leisure; and he had already developed a strong inclination towards science. Accordingly, he continued his scientific studies. He was fortunate to have some very remarkable teachers: the astronomer and mathematician Lacaille, the botanist de Jussieu, the geologist Guettard, and the chemist Rouelle, four outstanding eighteenth-century French scientists. From his contact with them he derived much of the inspiration that characterized his own original work in later years.

The Abbé Nicolas Louis de Lacaille (1713-62) had been an assistant at the Paris Observatory before accepting the poorly paid post of Professor of Mathematics in the Mazarin College, where he had installed a small observatory. He was a tireless worker and his death at the early age of forty-nine was attributed to overwork. Lacaille had spent four years, from 1750 to 1754, at the Cape of Good Hope in charge of a very successful expedition despatched there by the Paris Academy of Sciences. Under his direction the expedition measured an arc of the meridian, that is, they measured the distance along the meridian, the line running north and south, between two places of known latitude, and thence calculated the circumference of the earth. Many other observations were made; and many new stars were observed and catalogued, a great advance being made on the similar work that had been carried out by the English astronomer, Edmond Halley (1656-1742).* Despite the inadequacy of his equipment at the Mazarin College, his lack of assistance, and the necessity of earning his living by other means, Lacaille

* Applying the mechanics of Newton, Halley predicted that the bright comet seen in 1682 and at earlier dates would return at the end of 1758 or the beginning of 1759. It was observed on Christmas Day, 1758. This remarkable visitor from the confines of our universe appears at intervals of about three-quarters of a century and has since become known as Halley's Comet. Halley's successful prediction of the date of its return was a striking confirmation of Newton's mechanics.

won high reputation as a skilled astronomical observer. He taught Lavoisier mathematics and astronomy.

In botany Lavoisier's teacher was Bernard de Jussieu (1699-1777), who had begun to apply the classification of Linnaeus to the plants in the garden of the Trianon. De Jussieu did not publish his work, but he was a pioneer in his subject and his researches were continued by his better known son, Antoine Laurent (1748-1836).

In mineralogy and geology Lavoisier was taught by Jean Étienne Guettard (1715-86), one of the founders of the science of geology, and the pioneer of geological surveys and geological maps. Guettard was a remarkable person; his manner was brusque and even hot-tempered towards those in power and authority, but he was kind to his young contemporaries in whose company he is said to have been gay and lighthearted. His temperament is revealed in his reply to one who thanked him for support in election to the Royal Academy of Sciences: "You owe me nothing," said Guettard, "for if I had not thought it just to give you my vote, you would not have got it, because I do not like you."

Lavoisier studied chemistry under Guillaume François Rouelle (1703-70), demonstrator in chemistry at the Jardin du Roi from 1742 to 1768, and renowned throughout the scientific Europe of his day for his precise presentation of chemical facts free from the theoretical obscurities in which other expositors often enveloped them. Rouelle is remembered in chemistry for his classification of salts according to their crystalline form, and also as the teacher of many of the French chemists of that time, not only of Lavoisier, but also of Cadet, Macquer, and Bayen. He was greatly admired by Diderot, editor of the French *Encyclopédie,* who, with many other notable figures from all parts of Europe, was an enthusiastic frequenter of his lectures. And what memorable lectures they were! And what an audience!

According to long-established custom the lectures at the Jardin du Roi were given jointly by the professor and the demonstrator. The professor began each discourse and dealt with chemical theory, scorning to lower himself to the base drudgery of experiment or to soil his hands with the charcoal used as fuel in the chemical furnaces. Then the demonstrator followed. His duty was to provide convincing experimental evidence and proofs in support of the general ideas and theories that the professor had just been expounding. In Rouelle's day, the professor was Bourdelin, whose theoretical disquisitions fell coldly upon his hearers; each time as he ended he announced: "Such, gentlemen, are the principles and the theory of this operation, as the demonstrator will now prove to you by his experiments."

But the demonstrator was Rouelle: and more often than not his experiments proved the very opposite of what the professor had been teaching, and flatly contradicted his theories. The distinguished audience of Parisian scientists, authors, gentlemen of the Court, ladies of fashion, students, and others followed the demonstrator's experiments with all attention, for Rouelle's demonstrations attracted the intelligentsia of Paris and popularized the science of chemistry in eighteenth-century France.

Rouelle arrived in the theatre of the Jardin du Roi in full dress. He began calmly, then warmed to his work. As his enthusiasm grew, his thought seemed to delay him and he became impatient. Off went his hat—which he sometimes hung on some part of the chemical apparatus. Then, in turn, off came his powdered wig; he untied his cravat; and, without a pause in his speech, he unbuttoned his velvet coat and his waistcoat and took them off, too.

In the experiments Rouelle was helped by his nephew, but the young man must have found that helping uncle was neither the easiest nor the lightest of duties. Just as the audience could not have forecast where Rouelle's thoughts or his restless limbs would next lead him, so the unfortunate nephew found himself in equal uncertainty, and he never appeared to be where he was wanted or ready with the apparatus that his uncle might be demanding. So Rouelle, without halting in his lecture, and as if his audience were still before him, would go to some back room or other of the laboratory to find the necessary apparatus. When he came back to the theater, his demonstration was, of course, usually completed and he would re-appear with the words: "Yes, gentlemen, that is what I had to say to you." If he were asked to repeat the demonstration, he did so willingly, thinking that he had not been well understood. In his high excitement, he often gave unconscious expression to new ideas or he revealed processes that he would have preferred to conceal, and then he would say: "That is one of my secrets that I tell to nobody."

Many stories are told of this lovable character. His complaints against his contemporaries seem to have been an essential part of his lectures. Macquer, Buffon, Bordeu, and others were named, but all remained his friends. His favorite epithet was "plagiarist." He even accused the unfortunate French commander defeated at Rossbach of being a plagiarist. Buffon pointed out to him that it was not a plagiarism to allow oneself to be beaten by the Prussians, but an entirely new invention of the French general, the Prince de Soubise. Buffon's triumph was momentary. "Do not defend him," retorted Rouelle, "he is the lowest of animals, a horned mule, a downright blind pig. I am certain that he has something foul in his make-up!"

The Prince de Soubise had been defeated, and crushingly defeated, by Frederick II of Prussia. In Paris, Rouelle said that he felt as if he had been ground up and that all the Prussian cavalry had ridden over his body; in London, Whitefield, the Methodist preacher, hailed in his Tabernacle a victory for Protestantism.

Rouelle was a patriot, upright, honest, generous, incorruptible, and an enemy of all meaningless verbiage in science. Lavoisier was fortunate in obtaining a copy of Diderot's notes of Rouelle's lectures and meditated on the contents, setting down further brief notes and reflections of his own about substances and elements.

During this time Lavoisier extended his scientific studies to anatomy, thereby acquiring the knowledge that was later to help him in his physiological researches. But as yet, he was only a beginner feeling his way. He was now a young man of twenty and he was attracted towards mathematics and also towards meteorology, which, in fact, became one of his lifelong interests. He began to record several times a day at his home in the Rue du Four-St.-Eustache the readings of the barometer, a practice which he persuaded his relatives to follow when he was away from home at any time. Later, when he had acquired a scientific reputation, he recorded such observations in other parts of France for comparison. With the aim of discovering the laws relating to the movement of the atmosphere, he sent barometers to correspondents in various parts of France in order to get suitable data, and also to others as far away as Aleppo and Bagdad. With sufficient data of this kind he considered that it would be possible to forecast the weather. He drew up in 1790, from observations extending over thirty years, a set of rules for this purpose, although he pointed out that a knowledge of the force and direction of the winds and of the degree of moisture contained in the air was also necessary for such predictions. He held that when these data could be obtained completely and regularly, it would be possible to predict the weather for one or two days ahead with reasonable accuracy, and to publish these forecasts in some kind of daily journal, which would be very useful to society. Only his untimely death in 1794 prevented the further prosecution of this interesting and valuable work, which has since been carried to completion by others. But, for convenience, we have anticipated the future a little at this point and we must return to the year 1763.

Already, Lavoisier had revealed something of the breadth of his mind and his insatiable zest for work, for which he put aside all other distractions. At nineteen, in order to escape social duties, he made the excuse of ill health and put himself on an exclusive diet of milk for some months. His friends were convinced that he was unwell. One of them, de Troncq, sent him some gruel, accompanied by the advice: "Your health, my dear mathematician, is like that of all literary men, whose minds are stronger than their bodies: so spare your studies and believe that a year longer on earth is worth more than a century in the memory of men."

By the age of twenty, therefore, Lavoisier had had a good education and had studied law, which meant that he had studied the classics (even now probably the best preliminary discipline for scientific studies). He had learned to write in his own native language of French in the clear and logical style that later marked his scientific memoirs and books. He had had great teachers, most of whom were actively engaged in original work, in research, as we would now say, in the increasing of knowledge. He had been in close personal contact with them all, botanizing with de Jussieu and accompanying Guettard on geological excursions, making chemical experiments in Rouelle's laboratory and astronomical observations with Lacaille in the small observatory at the Mazarin College. A broad education, based on the study of the classics, literature, and law, followed by scientific studies under such teachers, made an ideal preparation for the development of his powerful mind. In France universities still require graduates to write an essay, to write in their own language, and to do so as part of their education. Lavoisier's scientific writings provide us with good evidence of the wisdom of this rule. Moreover, we see him at twenty, studying widely over a range of subjects and not specializing in his early 'teens, a practice that is now unhappily reversed in many countries.

As for his life at home, he lived with his father and his aunt in the quiet house in the Rue du Four-St.-Eustache, to which Mme. Punctis admitted only great intimates, among whom was the redoubtable Guettard.

III

FRANCE BEFORE THE REVOLUTION

THE CENTURY INTO WHICH LAVOISIER HAD BEEN BORN became one of the most eventful in the history of France. His brief life of fifty-one years fell into the more stirring half of that age. Unremedied social discontent was eventually followed by revolution and revolution by military despotism. Not the least important cause of that memorable upheaval was the unsound fiscal system which France had had throughout her years. Her people would not pay taxes, and her statesmen apparently did not know how to impose them justly. From the fourteenth century exemption from taxation had been customarily granted to privileged classes, such as the nobles and the holders of certain offices obtainable by purchase. Collection of taxes was "farmed out" to a company of speculating tax-gatherers, called the *Ferme Générale* or Tax Farm, who were hated as robbers by the people whom they oppressed. Frenchmen tried to avoid the exactions of these officials by all possible means; and, as most taxable Frenchmen were poor, France eventually became, in Calonne's despairing words, a country impossible to govern.

The Tax Farm was a company of financiers who bought from the Government, as an investment of capital, the privilege of collecting certain indirect national taxes and other customs and duties, under leases which were renewable for periods of six years by mutual agreement with the Minister of Finance. The Government required a stipulated annual revenue from these sources and the Farmers, according to their agreement, undertook to provide this by a sum paid in advance in each year of the lease. The taxes might yield more or less than the agreed amount and, of course, the collection had to be financed and organized efficiently. The greater the yield, the more the Farmers stood to gain. They received no direct payment from the Government for their work; they collected the taxes as a business venture with the chance of profit or the risk of loss, and they generally took care that there was no loss.

The Farm in Lavoisier's time owed its general form and organization to Jean-Baptiste Colbert (1619-83), Louis XIV's Minister of Finance. Before his reforms collection of the indirect taxes had been farmed to several companies of financiers, who had enriched themselves at the expense of the King, and to the detriment of his subjects. To remedy these abuses, Colbert in 1681 effected a change by which the taxes were farmed to one company only, and for a period of six years with renewable leases. Although this reform did not eliminate malfeasance, the administration of the farming system had been much more efficient than that of the Government; and Colbert had good reason to hope that unification would lead to greater honesty and stability, and that the Farmers would play the part of the nation's bankers. The arrangement persisted with various changes until the Revolution.

The King's Farmers-General, as they were called, collected taxes on salt and tobacco, customs duties on drink, import duties levied at the national frontier and at the boundaries between the provinces of France, the tax of 3 per cent on merchandise imported from America, and the customs duties and tolls for entry into Paris.

Shortly before a lease expired, terms for its renewal were agreed upon by the Minister or Controller-General of Finance and the Farmers-General after lengthy discussions on the recent returns of revenue. This done, a new lease was drawn up, not in the name of the Farmers-General, but in that of a man of straw, the Contractor-General, with the Farmers-General named as his sureties. The Contractor then disappeared from the scene, although all business was done in his name. He did not go unrewarded, however, for he was granted an annual pension of 4,000 livres without responsibilities or duties during the term of the lease. This device enabled the Minister to provide for one of his followers the sum of 24,000 livres over a period of six years at the expense of the Farm, which meant in reality at the expense of the taxpayers.

When the Abbé Terray was Controller-General of Finance, fifty-five of the sixty places in the Farm were similarly burdened with the provision of emoluments for favorites and other parasites; two of the remaining five Farmers-General were said to have bought their freedom from this exaction directly in hard cash. Identity of the beneficiaries was, of course, a carefully guarded secret; but, when the lease of Alaterre was about to expire in 1774, Terray wanted detailed information about the sums involved and he wrote a confidential circular letter to the

Farmers-General. By the indiscretion of a clerk the list was published and Paris learned that pensions of this kind amounted to 400,000 livres annually, and that the incomes of fourteen places in the Farm went to courtiers and favorites, indeed, even to royalty. The King, it appeared, received the entire income from one seat in the Farm; the Dauphin and sisters and aunts of the King had 50,000 livres; and there were smaller sums for the less august.

Some of the Farmers-General had offended public opinion by the luxury of their lives, but many had been upright administrators. It was later shown, in fact, that the fortunes spent in extravagance by some former members of the company could not have been acquired from their places in it, but must have been amassed by other speculations. Public opinion, however, inevitably connected their wealth with the exactions of the hated Farm. Michel Bouret, one of three brothers forced on the Farm by the influence of Mme. Pompadour, had wasted over forty-two million livres in riotous opulence; he had gained this fortune, however, not in the Farm, but by speculations in wheat. There were other such instances. There were also financiers not in the Farm whose luxury was offensive to public feeling. Beaujon spent 200,000 livres a year on his "cradle-rockers," as Paris named the company of pretty girls who came in the evenings to tell him bedtime stories and to lull him to sleep with songs. Financiers had an evil reputation in Paris; and to Paris the Farmers-General were nothing if they were not financiers.

This obsolete system continued. Despite Colbert's contention that the real greatness of a country was to be measured by its wealth, and that its wealth was produced by the work of its people and not by splendid uniforms and glittering Court functions, he failed to introduce an equitable method of taxation. In fact, to provide supplies for a King who waged war for forty years, Colbert had to increase the burden on the people. Smugglers, who tried to evade the infamous *gabelle* or salt-tax, were hunted with bloodhounds; collectors were punished for inadequate exactions.

Yet the nobles remained untaxed, although under the new centralized rule of Louis they had become mere courtiers at Versailles, not unruly barons acting as a useful check on the absolutism of the Crown. Moreover, as if to hamper trade, customs duties were imposed between provinces; offices with the privilege of exemption from taxation were still sold; and the wasteful method of collection remained unaltered. Taxes fell on the poorest, and exemptions were too well established to be abolished. After Colbert, the finances went from bad to worse.

The States-General, representing the three classes of France, had not met since 1614; and Louis XIV had refused to summon it. When it had met, its concern had been with its own selfish interests and not with the welfare of France. Such opposition as there was to the Court came from the Parliament of Paris. This was in no sense a representative body, however, merely a corporation of lawyers whose membership was either inherited or bought. To some extent it checked the Crown, but even though its members were critics of the Court, they were lawyers and opposed to fiscal changes. The weary taxpayers of France could not expect to find any champions among these officials.

Louis XIV ended his long reign of seventy-two years in 1715. Under his successor, Louis XV, who reigned for fifty-nine years until 1774, discontent increased. The new King was not popular; the worldly priests were despised; the Farm was hated. Internal quarrels flourished. Frenchmen were struggling with one another for political freedom and fiscal reform. Lawyers of the Parliament of Paris were ranged against the political writers who advocated change.

At the turn of the half-century France entered upon the disastrous Seven Years' War (1756-63), in which she lost Canada and her Indian Empire. At sea, her Mediterranean and Atlantic fleets went down to disaster, the latter in the great action fought by Hawke in Quiberon Bay in 1759. Pitt, the architect of British policy, had destroyed French power in America and in India. France had lost her great oversea possessions because of her internal and Continental preoccupations; and the support that she later gave to the American colonists in their fight for independence against Britain proved the fatal blow to her finances.

Yet, while France and Britain were military enemies throughout the century, British influence on French thought was considerable. Many Frenchmen respected the political system across the Channel, where their neighbors had carried out a bloodless revolution with remarkable efficiency in 1688 and had successfully led the resistance to Louis XIV. Among the islanders there seemed to be a freedom that was beyond the reach of French statesmanship.

French thought had been influenced by Locke (1632-1704) and Newton (1642-1727). Locke had advocated religious toleration; he had disputed the theory of the divine right of kings; and he had taught that supreme power lay in the hands of the people. Newton's discoveries were the subject of a considerable popular French literature. Voltaire (1694-1778) did much to make the work of this greatest scientist of all time known to French readers. During his three years of exile from France (1726-

29), following a quarrel with a nobleman, and incarceration, without trial, in the Bastille, he had found England a pleasant refuge. He saw Newton's funeral procession; and he observed that noblemen, the Lord High Chancellor of England, the Dukes of Montrose and Roxburgh, and the Earls of Pembroke, Sussex, and Macclesfield, accompanied the body to its last resting-place in Westminster Abbey. Obviously, the English had a different scale of values from that applied when Voltaire had been flung into the Bastille. Tyranny and inequality appeared to be unknown; speech was free; a subject could not be cast into prison without trial; nobles and priests paid taxes; religious differences were tolerated and a man might go to Heaven along a road of his own choosing. In his *Letters on the English* (1733), Voltaire related to his countrymen his experiences of English life.

Later Montesquieu (1689-1755) visited England and extolled its extraordinary degree of political freedom. "England," he said, "is the freest country in the world." In his book *The Spirit of Laws* (1748), he studied the various political systems and criticized the abuses of the French monarchy.

Voltaire and Montesquieu had brought about much admiration in France for the English way of life, at least of political life, although they had not seen it in all its aspects. Their writings had a deep influence on French political thought and on French thought in general. While the old system staggered on to its ruin, enlightened thinkers of France, the *philosophes,* produced from 1751 to 1776, under the editorship of Diderot (1713-84) and d'Alembert (1717-83), the 35-volume *Encyclopédie,* in which they challenged the obsolete, the unjust, and the superstitious in European society. For all these evils, the remedy was liberty of thought and speech; with these liberties won, human happiness seemed attainable and human nature perfectible. It was an optimistic philosophy, especially the belief in the perfectibility of human nature, which is not yet entirely rejected in some circles, although much more likely to be doubted than it once was.

In the field of economics, under Quesnay (1694-1774) and more particularly under Turgot (1727-81), there arose a school called the Physiocrats, who regarded the land as the only source of a country's wealth, and free and unrestricted trade as essential to national prosperity. This doctrine of free trade passed to the Scots economist, Adam Smith (1723-90), with historic results on British economic policy. Adam Smith had visited Voltaire in France and had been associated with the physiocrats. In 1776 he published his classic *Inquiry into the Nature and Causes of the Wealth of Nations.* Unlike the physiocrats, he held that labor, not land, was the source of wealth, but he shared their belief that commerce and industry should be free.

The physiocrats were practical men. To their voices was added another of a different type, that of Jean-Jacques Rousseau (1712-78), whose *Social Contract* (1762) pictured a state where power resided in the people and where every citizen was trained to virtue, that is, to do unto others as he would have others do to him.

In France there was arising a middle class, eager and intelligent, critical and competent. During the eighteenth century French trade had increased five-fold. But the middle classes found themselves restricted in a social fabric that was medieval, feudal, and rotted by privilege; and they began to question the rightness of the system and to criticize it and to write pamphlets attacking it. They paid taxes; the nobles paid none, nor did the clergy. The former, in general, rendered no service to their country and the latter were, in the main, wealthy and worldly, to say no worse. There were, however, nobles who still lived on their estates and cared for their people, and priests who had not forgotten that their kingdom was not of this world.

Such was the France in which Lavoisier grew to manhood and in which he was destined to play a great part, not only as a scientist, but also as an administrator and public servant; and such were the social problems that he presently heard debated in the salons of the class into which he had been born.

IV

LAVOISIER'S CHEMICAL HERITAGE

TO UNDERSTAND AND TO APPRECIATE THE ACHIEVEMENTS BY which Lavoisier in his later years transformed chemistry into one of the great sciences, rivaling mathematics, mechanics, astronomy, and physics in its dignity and importance, helping man to a better understanding of the world around him, providing lavishly for his daily needs, and contributing greatly to the progress of his civilization, we need to devote some space to consider the chemical *milieu* into which Lavoisier was born, before detailing the steps by which he slowly effected this remarkable revolution and did for chemistry what Newton had already done for mechanics.

The four elements

WHILE ASTRONOMY IN ITS REMOTE BEGINNINGS IN THE OBSERVATION of the motions of the sun, moon, and stars, is said to have been the earliest of the sciences, primitive chemistry may well rival it in antiquity. The men of the early civilizations can scarcely be thought to have fixed their gaze on the heavens and disregarded the useful properties of the substances lying at their feet.

Some kind of practical chemical knowledge was common to all the early peoples. The arts of cooking, baking, brewing, pottery, and metal-working, are all really chemical in their nature. In Babylon and in Egypt these arts and crafts, including the making and coloring of glass, were so highly developed that they excite our admiration today in the specimens that have survived to our time.

So far as we know, however, the first philosophers to think about and to try to explain the properties of substances and the changes that they undergo were the Greeks of the sixth to fourth centuries B.C. Thales in the sixth century considered that all substances were formed from water. Empedokles in the fifth century thought that they were produced by the combination of different proportions of four elements, earth, air, fire, and water. In the fourth century Aristotle taught that there was a prime or first or simplest matter from which all substances, even the four elements, were ultimately derived. Speculation of this kind was premature and there were not enough facts available to the Greek thinkers for the formulation of a more satisfactory theory. Besides, they had not yet learned the controlling part that experiment must play in the advance of science and the discipline that it must be allowed to exert on scientific imagination, which is mere daydreaming and phantasy unless it is constantly subjected to the arbitrament of experimental test. And it was a handicap to them that their social system led them to look upon practical knowledge, such as that of the chemical arts and crafts practised in the workshops of their cities, as debasing and beneath the dignity of study by philosophers. They analyzed Nature with their minds only, not by experiments as well.

Under the practical-minded Romans, theoretical science languished for lack of attention. Later, when the philosophers of Rome and Byzantium were forced by political changes to seek refuge in Alexandria, that cosmopolitan city open to all the known world, the speculative powers of the Greek spirit at last made direct contact with a vast body of practical chemical knowledge which had been handed down through the centuries from ancient Egypt. Fact and theory met in Alexandria on some sort of equality, and here the science of chemistry first arose in the early centuries of the Christian era.

Since all matter was thought to be composed of the four elements combined in varying proportions, the Alexandrian chemists concluded that all substances were convertible and transmutable, one into another, by suitably adjusting the relative proportions of the four elements of which every substance was formed. Their attempts to effect what their theory predicted failed; and so they made a further supposition, namely, that a special substance, which later came to be known as the Philosophers' Stone, was necessary to achieve transmutation. They sought for the Stone and thus began the long history of alchemy, too often remembered for only one of its aims, the transmutation of base into noble metals, of lead into silver and gold. Parallel with this arose a belief in an Elixir of Life, which would cure all the ills that flesh is heir to; and this also was sought for. Both searches failed, and belief in their possibilities became discredited and eventually regarded as absurd.

Today, transmutation does not sound so absurd as it did even fifty years ago, but the progress of discovery in atomic physics has been swift in our time and the first transmutation ever achieved (1919) was effected through our knowledge of the structure and components of atoms. The ancient approach was, of course, very different from the modern. The ancient theory was founded on inadequate fact, but it was not absurd, since there were many observations to support it. When water was boiled and evaporated, some earthy matter was left; and it seemed that water was, at least partly, convertible into earth. This belief was not finally rejected until Lavoisier, as we shall see (Chap. VII), disproved it experimentally in 1770, which is as much as to say that chemists abandoned it only about 180 years ago, a mere yesterday in human history.

Another observation made by early chemists was that lead when heated for a long time was replaced by a small amount of silver. They concluded that the lead had been partly changed into silver, because they did not know that the silver was already contained in the lead. Many lead ores contain silver, and in modern times the recovery of this silver and its separation from the lead has become an important industrial process. In those earlier times, however, it was thought that the silver had been produced by a change in the lead, and that, if it could be discovered how this had happened, similar changes could be effected in other substances. Why, for instance, could not base lead be transmuted, not partly into the noble metal silver, but completely into the noblest and most precious metal, gold?

Iron when put into a solution of blue-stone or copper sulphate disappeared; and copper appeared in its stead. Thus the simple and now familiar experiment of coating a steel knife blade or pen nib with copper by immersion in a solution of copper sulphate would have appeared to the alchemists as a transmutation of iron into copper, because they did not know that the copper in the solution had been replaced by the iron, that the metals had changed places.

Many facts of this kind were known to the Alexandrian chemists. It is not surprising that they believed that substances were transmutable; it would have been surprising if they had not believed it.

The new science, already named *chemeia,* passed with political supremacy to the Muslims, who greatly developed it under the name of *al-kimia.* When Muslim culture entered Europe, the name became *alchemia* or alchemy, whence we get *chymistry* as the English word, and later *chemistry*. During the thirteenth, fourteenth, and fifteenth centuries, the belief that substances were transmutable dominated the chemical thought of western Europe.

Paracelsus (1493-1541) criticized the chemists of his age in very harsh terms for devoting their labors to the search for the Philosophers' Stone. He called upon them to seek out new medicines, especially mineral remedies. As a result, chemistry became for a time the mere handmaid of medicine. Paracelsus also taught that, while the four elements were the ultimate inmost constituents of all substances, there were three more immediate constituents, the *tria prima* or three principles or elements, namely, salt, sulphur, and mercury, into which substances might actually be decomposed. These, of course, were not the common substances usually known by the names of salt, sulphur, and mercury, but principles or elements resembling those familiar substances in their general and characteristic properties.

A change from a world composed of four elements to one made up of three is not very significant, except as it represents the first break with a theory accepted for nearly two thousand years.

Van Helmont's gas

FOLLOWING PARACELSUS CAME JOHANN BAPTISTA VAN HELmont (*c.* 1580-1644), who revived the ancient theory of Thales that all substances were formed from water. Van Helmont, however, excepted air. He introduced the term "gas," and he was able to recognize in a simple way that one gas differed from another, but he did not devise any method for collecting gases.

Chemists had long been familiar with the explosions produced in closed vessels when chemical changes shattered the vessels. Van Helmont said that these explosions were due to the production of gas. Among the different gases, he noted one that was produced when charcoal was burned; and there was a poisonous gas in the mines, which extinguished the flame of a candle, and also in the Grotto del Cane, or Dogs' Grotto, near Naples, into which a man might safely walk, breathing the wholesome air at the higher level, while a dog fell dead instantly, asphyxiated by the lower layer of poisonous gas. A gas was also produced in the fermentation of grapes in the making of wine. All these gases were, we now know, the same, namely, carbon dioxide or carbonic acid gas, which is formed in the burning of wood and coal and in our own respiration, is present in mineral waters, and is used to make "soda water."

Van Helmont detected a totally different kind of gas produced in the intestine and in putrefaction. It would take fire and burn and he called it *gas pingue,* or combustible gas, to distinguish it from the other kind, to which he gave the name *gas sylvestre,* or incoercible gas, because it shattered to pieces closed glass vessels in which it was produced. He could not, of course, make very clear distinctions between the different kinds of gases and he sometimes confused them, but he was the first to demonstrate that such aerial substances existed and that they were material bodies. These discoveries were published in 1648 in his posthumous *Ortus Medicinæ,* or *Dawn of Medicine,* but no gas had been isolated.

A simple method for the collection of gases was devised by the Hon. Robert Boyle (1627-91) and used by him in 1659. It is shown here in Fig. 5. The glass vessel A, filled with "Oyl of Vitriol and fair water, of each almost

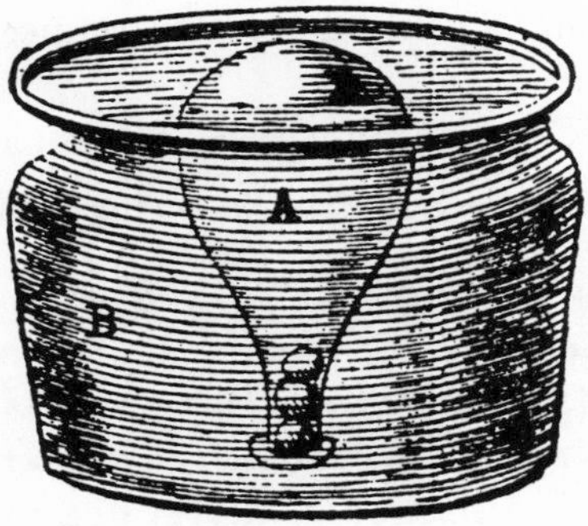

5. The first apparatus devised for the collection of gases. Robert Boyle, 1659.

a like quantity" (that is to say, dilute sulphuric acid), was inverted over water in the vessel B. By "casting in half a dozen small Iron Nails" into A before inverting it into B, Boyle had a reacting mixture which gave the gas we now call hydrogen, which rose in bubbles to the top of the vessel A, displacing liquid which passed out into B. When he used *aqua fortis* or nitric acid instead of oil of vitriol, he obtained the gas now known as nitric oxide. He regarded both these products, however, not as two different gases, but as air regenerated from the two reacting substances, the acid and the metal in each case. He supposed that all matter consisted of minute particles capable of rearrangement, so that those forming a liquid, or even a solid, might be rearranged and rarefied to give air. For more than a century after van Helmont, chemists followed Boyle in thinking of air as the only substance of its kind, and of such gases as they met with as mere varieties of common air produced by some slight change in its constitution, but still essentially air. Boyle, however, had successfully devised a simple apparatus for the collection of the "air" given off from substances by chemical reaction.

Boyle defines the chemical "element" (1661)

IN HIS "SCEPTICAL CHYMIST" (LONDON, 1661) BOYLE launched a criticism of the current beliefs in the theories of the four elements and the three principles, one or other of which was generally accepted by the individual chemist. He especially attacked, and refuted by many facts drawn from well-known chemical observations, the underlying assumption of these theories, namely, that fire resolved substances into their component elements. To this end, he pointed out that the nature and number of the products obtained from a given substance depended on the manner in which the fire was applied, that is, on whether the substance was heated "in an open fire" or "distilled in a retort," that is, subjected to combustion or distillation. He reminded chemists that wood when burned was "sequestred into ashes and soot," into two products, but when it was subjected to distillation it yielded oil, spirit, vinegar, water, and charcoal, five products.

So, he argued, by the action of fire one and the same substance may be decomposed into either less than three or more than four components, which was against both theories. Clearly, "the operation of fire may be diversified by circumstances" and fire is not the genuine instrument by which substances are decomposed into their elements. Similarly, in an instance where one knew the components beforehand, as in the preparation of soap, which is made from fat and alkali, fire did not decompose soap into the substances from which it had been made, but into an evil-smelling oil and a saline earthy substance, both useless for making soap. The products of the action of fire on substances were, said Boyle, not elements, but new compounds formed by the rearrangement of the minute component particles of the substances through the action of the fire upon them. Chemists, he urged, must think more closely and fit their theory to the facts. To clear the way, he defined the chemical elements in historic words:

> *I mean by Elements, as those Chymists that speak plainest do by their Principles, certain Primitive and Simple, or perfectly unmingled bodies; which not being made of any other bodies, or of one another, are the Ingredients of which all those call'd perfectly mixt* [*i.e. compound*] *Bodies are immediately compounded and into which they are ultimately resolved.*

Stripped of its seventeenth-century dress, this means that "an element is a substance that cannot be decomposed into anything simpler." Boyle was, unfortunately, unable to apply his definition by drawing up a list of chemical elements or even by indicating any one element; but, substantially, as we shall see, Boyle's definition was used by Lavoisier in 1789 when he drew up the first table of chemical elements as "substances that have not been decomposed."

Boyle, Hooke, and Mayow on combustion and respiration

BOYLE STUDIED, ALSO, THE PROBLEMS OF COMBUSTION AND respiration. An air pump had been constructed for him by Robert Hooke and in 1660 he published an account of various experiments made in a large receiver A attached to the pump (represented in Fig. 6). Through the tap S connection was made with the air pump and the receiver could be evacuated. A stopper BCDE, which could be made airtight by cement, enabled apparatus to be placed inside A for experiments. Boyle found that flames were extinguished and animals perished, when enclosed in the receiver, much more rapidly on evacuation of the air than if the air were not removed. He knew, of course, that in the open air flames lasted very much longer and animals continued to live; and he concluded that air supported both fire and life in some way not yet understood.

Hooke, however, in his *Micrographia* of 1665 went much further. He asserted that there was a substance in the air, which was also present in saltpeter or niter, and that it was this substance that was active in combustion. Later he said that this same substance formed the part of the air that served for respiration. Hooke's most important evidence was that mixtures such as gunpowder, which contained saltpeter, would burn under water, that is, in the absence of air. Hooke's substance that was present in air and in saltpeter we now know to be oxygen. It was a bold guess, but not supported by enough experimental evidence.

Then in 1670 Boyle made further experiments on respiration. He put a mouse and a pressure gauge into a closed glass vessel; and he found that, despite the respiration of the mouse, the pressure of the air did not change. If air were consumed, a decrease of pressure, he thought, would

6. The glass receiver in which Boyle studied combustion and respiration.

have been observed. But, as we shall see, the demonstration of this required an experiment of a different kind. In 1672-73 Hooke claimed before the Royal Society of

London that he had found that air was diminished when combustion took place in it, but he was unable, despite several attempts, to reproduce this important experimental result and he did not state what apparatus and arrangement he used.

7. Mayow in 1674 showed that a lighted candle floating on water consumed part of the air in a cupping-glass inverted over it; the flame went out and the water rose inside the cupping-glass. A similar result was given when a piece of camphor was ignited from outside by means of a burning-lens.

Boyle concluded in 1674 that air contained a vital substance necessary for the maintenance of fire and of life; and, since in his experiments the air did not appear to be diminished by these processes, the proportion of this vital substance in the air must be very small. Both Boyle and Hooke had failed to establish experimentally that air was consumed in combustion and respiration.

In this same year of 1674, however, by a change in the method of experiment, John Mayow (1641-79) established the fact for both processes. He showed that a burning candle and kindled camphor both consumed air; and that a mouse consumed air in breathing. He made his experiments over water, in which, of course, the gaseous product of combustion and respiration is soluble. Boyle's method of experiment in the receiver of his air pump or

8. Mayow in 1674 showed that part of the air was consumed in respiration. The lower vessel was covered with a piece of moistened bladder and a mouse was placed in the upper vessel resting on it. The bulging of the bladder into the upper vessel showed that part of the air had been consumed in respiration, which could be further demonstrated, when the lower vessel was not heavy, because it was then possible to lift the whole apparatus without its coming apart by grasping only the upper vessel in the hand.

in a closed glass vessel, however, could not reveal that the air had decreased in volume. Mayow's experiments are represented in Figs. 7 and 8.

Mayow concluded from his results that air contained "nitro-aerial particles" or "nitro-aerial spirit," which formed a part of the air and which was essential for combustion and respiration. His quantitative results were not good; he found a diminution of air of only one-thirtieth by combustion and one-fourteenth by respiration. He had, however, established the fact of these decreases and he was the first to do so. It is unfortunate that he applied his theory too widely and introduced many contradictions into it, for which he has since been justly criticized. It will be observed that Mayow's apparatus (Fig. 7) bears some resemblance to that used in 1659 by Boyle (shown in Fig. 5). In fact, it was Mayow in 1674 who gave the first illustration (Fig. 5) of Boyle's apparatus. The change to experimentation over water had brought to light this most significant result, the consumption of a part of the air both in combustion and in respiration. The problem was, however, not carried any further at this time.

The calcination of metals

COMBUSTION AND RESPIRATION WERE, THEREFORE, COMING to be recognized as processes in which air or some substance in the air took part. There was also another process in which we now know that part of the air is concerned, namely, the calcination of metals. When, for instance, lead is heated in air, it loses its metallic properties and appearance and is converted into a powder, known in earlier times as a calx. Tin, copper, iron, and other metals, are similarly convertible into calces. Rust is a calx of iron, but is produced without the action of heat. It was known that in this change or calcination the metal gained in weight, since the calx weighed more than the metal from which it had been produced. Today, we recognize that the gain in weight is due to the combination of the metal with the oxygen in the air.

Boyle published in 1673 in his *Essays of Effluviums* the results of many careful experiments that he had made on the calcination of metals. He heated weighed amounts of various metals in open glass vessels and in crucibles and then weighed the calces produced. The metals had gained in weight. In some cases, he heated a weighed amount of metal in a sealed glass vessel until the metal was calcined; then he opened the vessel, removed the calx and weighed it, and found that it had increased in weight. He considered that in all these cases the gain in weight was produced by material particles of fire, which had passed through the crucibles and glass vessels, and had been taken up by the metal in its conversion into calx.

In 1674 Mayow offered another explanation: the metals had gained in weight because they had taken up "nitro-aerial particles" from the air. This explanation is nearer to that later put forward by Lavoisier and which we now accept.

The "phlogiston" theory

MEANWHILE, THERE AROSE A NEW THEORY OF COMBUSTION, the "phlogiston" theory, first formulated by a German chemist, Johann Joachim Becher (1635-82), and then greatly developed and extended by another German chemist, Georg Ernst Stahl (1660-1734). According to this theory, all combustible bodies contained a common inflammable principle or element of fire, which Stahl called "phlogiston." When a combustible was burned, its phlogiston was given off and escaped. This seemed a very reasonable theory. Inflammable substances, at first sight, appear to contain something in common, something that escapes from them all, as fire and its concomitants, flame and light, when they are burned.

Thus an inflammable substance lost its phlogiston when it was burned. The more phlogiston it contained, the smaller the residue left on its combustion. Such substances as oils, fats, and charcoal, which burn away almost completely, with little or no residue, must, therefore, consist largely of phlogiston.

It was known that, when the calx of a metal was heated with charcoal, the calx was reconverted into metal and the charcoal disappeared. Hence it was concluded that the phlogiston of the charcoal had passed to the calx and thereby reconstituted the metal; and it followed that a metal was a substance consisting of its calx united to phlogiston and that metals were to be classed as combustibles.

Combustion and calcination were therefore regarded as processes of decomposition. In the calcination of a metal, the metal was said to have lost its phlogiston, leaving as a residue its calx, and the composition of the metal might be represented thus:

$$\text{Metal} = \text{Calx} + \text{Phlogiston}$$

It was, however, known that, when a metal was calcined, the weight of the residual calx was greater than the original weight of the metal, although according to theory the metal had lost its phlogiston, lost something of its substance. But how could the weight increase, when something material, namely phlogiston, had been lost from the substance of the metal? Gain in weight accompanying loss of substance was a contradiction of scientific and all other experience. Many attempts were made to resolve this apparent paradox. Some chemists were led to suppose that phlogiston was lighter than all other matter. Others alleged that it did not gravitate as other matter, but levitated, that it naturally rose upwards to the heavens, whereas all other substances naturally tended to fall to the earth, or, in other words, that it had a negative weight.

The phlogiston theory was widely accepted by the chemists of the eighteenth century because it explained so much in a simple way, and much of the more valuable work of the seventeenth century was forgotten, especially the researches of Boyle, Hooke, and Mayow, on combustion and respiration. The new theory explained so much because it was simple. Ultimately, it proved to be far simpler than Nature.

Joseph Black's "fixed air"

THE HISTORIAN OF CHEMISTRY, LOOKING BACK TO THE YEAR 1750, can see little of the promise of modern chemistry. In 1754, however, the first gleam of the dawn began to appear. In that year a young medical student at the age of twenty-six, Joseph Black (1728-99), submitted for the degree of M.D. in the University of Edinburgh his inaugural dissertation on the acid humor arising from food and on *magnesia alba* (white magnesia, used in medicine). A year later he read to a society in Edinburgh an expanded English version of this work, which was published in 1756 under the title of "Experiments upon Magnesia Alba, Quicklime, and some other Alcaline Substances."

Everything about this remarkable work is historic, even its origin. When Black was a student of medicine, there was much criticism of the use of caustic remedies as solvents for "the stone," or urinary calculus, because of their severe action on the tissues of the body. These remedies were often used and prescribed by quacks. Some years previously Walpole, the Prime Minister, and his brother Horace, who were both suffering from the stone, thought that they had been relieved by a remedy invented by a Mrs. Joanna Stephens. Moreover, they felt that such a valuable medicine should be made available to all and that Mrs. Stephens should receive her due reward. Accordingly, Mrs. Stephens indicated her willingness to accept £5,000, a considerable sum of money at that time, for her secret recipe, which was published in the *London Gazette* on June 19, 1739. It appeared that she used many ingredients, including calcined egg shells and snails, herbs, and soap. Her recipe is worth quoting:

> My medicines are a Powder, a Decoction, and Pills. The powder consists of Egg-shells and Snails, both calcined. The decoction is made by boiling some Herbs (together with a Ball, which consists of Soap, Swines'-Cresses, burnt to a Blackness, and Honey) in water. The Pills consist of Snails calcined, Wild Carrot seeds, Burdock seeds, Ashen Keys, Hips and Hawes, all burnt to a Blackness, Soap and Honey.

The further details stated that the powder was left to slake for two months, that six parts of egg shell should be put to one part of snail powder, that snails should be gathered only in May, June, July, or August, that the herbs for the decoction should be green chamomile, sweet fennel, parsley, and burdock, either as leaves or roots, and that the soap used must be the "best Alicant soap." This pharmaceutical blunderbuss was, in fact, fairly simple; it yielded, as we now know, a very caustic product, caustic potash, much more caustic than lime water. It is not surprising that medical opinion was opposed to its use. But Mrs. Stephens obtained her £5,000.

A new drug called *magnesia alba* or white magnesia had, however, been recently introduced into medicine. It resembled the mild alkalis such as chalk. Black began his research on *magnesia alba* with the object of finding out whether it would yield a remedy for the stone milder than the caustic alkalis.

Alkalis, mild and caustic, were and had long been important substances in chemistry and in industry, especially in the manufacture of soap. The mild alkalis, soda and potash, were converted into caustic alkalis by boiling with slaked lime, which was obtained by the familiar process of slaking quicklime with water, and quicklime had been produced for centuries by heating limestone or chalk in a kiln. It was thought that quicklime acquired its burning or caustic properties from some kind of fire-matter that was supposed to have entered into it when it was heated in the kiln; and that, when the mild alkalis, soda and potash, were boiled with slaked lime, this fire-matter was then transferred to them and thereby rendered them caustic.

Black first showed that, when a weighed amount of the mild alkali, *magnesia alba,* was heated, it lost weight. Because he could not collect any liquid product, he concluded that it had lost "air," which was known to be present in many substances. Moreover, the mild alkalis all effervesced on the addition of acids, which suggested that they contained "air"; but the caustic alkalis did not effervesce with acids. Therefore, the difference between the mild and caustic alkalis was that the former contained "air," while the latter did not. The caustic alkalis therefore did not derive their causticity from some fire-matter, but merely from the absence of this "air." Black followed out these changes by using the balance in a series of careful quantitative experiments.

Black called this "air" that he had detected, but not isolated, "fixed air," because at first he thought it to be common air "fixed" in the mild alkalis. He knew, however, that the surface of limewater, a caustic alkali, became covered with a white crust of chalk, a mild alkali, when exposed in shallow vessels for some time to the open air. He knew also that this crust did not appear, and that the limewater remained caustic, if it were put in loosely stoppered bottles, even when these contained some air. He concluded, therefore, that the "air" that turned the caustic alkalis into mild alkalis was not common air, but another species of air present in common air. This was a great stride forward; at last an "air," a gas, chemically different from common air, had been detected.

Later, Black showed that fixed air was produced in the burning of charcoal, in respiration, and in fermentation. His test was a simple one. He exposed the air produced in these three processes to limewater, a clear colorless liquid, and the limewater immediately became milky or chalky. Common air did not produce this change immediately, although it would do so after a long time, as there was a small proportion of fixed air in common air, as shown by the white crust of chalk, which we have just mentioned, formed on the surface of limewater exposed for some time to the common air in shallow vessels.

By way of confirming his conclusions about the production of fixed air in respiration, Black performed an extraordinary large-scale experiment, which has been reported as follows:

> In the winter 1764-5, Dr. Black rendered a considerable quantity of caustic fossil alkali [caustic soda] mild and crystalline, by causing it to filtre slowly by rags, in an apparatus which was placed above one of the spiracles in the ceiling of a church, in which a congregation of more than 1500 persons had continued near ten hours.

In other words, the fixed air or carbonic acid gas produced by the respiration of these unwitting subjects of Black's mass experiment had combined with the caustic soda filtering slowly through rags to give the mild alkali, soda. Black had been appointed Professor in Glasgow in 1756, after graduating in Edinburgh, and left to become Professor in Edinburgh in 1766. To the citizens of Glasgow, therefore, belongs the honor of having been the subjects of this striking experiment, conducted on some fifteen hundred of them, when engaged in a devotional exercise occupying the long period of ten hours.

Black's differentiation of an "air," or gas, from common air proved to be a turning-point towards the science of chemistry as we know it today.

Discovery of the gases

KNOWLEDGE OF GASES OTHER THAN BLACK'S FIXED AIR WAS still entirely lacking. Even the idea of a gas, of some aerial substance distinct from air, was new; for there had been no development of van Helmont's researches. And a new technique was required before chemists could isolate gaseous substances, indeed almost before they could discover them, although Black had recognized one "air" as distinct from common air without isolating it.

The new technique had already appeared, before Black's work, in its earliest form in the apparatus described in 1727 by the Rev. Stephen Hales (1677-1761), Vicar of Teddington, in his book on *Vegetable Staticks*. (The apparatus is shown in Fig. 9.) As will be seen, Hales had improved the simple apparatus of Boyle (Fig. 5), in which the "air" was collected in the vessel in which it was produced. Hales had separated production and collection; and his apparatus in the hands of Cavendish and his brilliant successor Priestley opened the way to a harvest of chemical discovery. Hales heated many substances and collected and measured the "air" released from them. He did not suspect that he had anything other than ordinary air as a product, and when he had meas-

9. Hales in 1727 devised the first "pneumatic trough," an apparatus for collecting gases over water. The gases evolved by heating substances in a gun-barrel, closed at one end, were led by the tube *rr* into the vessel *ab*, which was filled with water and suspended over water in the tub *xx*.

ured its volume, he emptied it out of the apparatus. He thought that air acted as a kind of cement binding together the minute particles of solid bodies. But he had devised the first "pneumatic trough," as this important and useful apparatus was subsequently named.

In 1766, ten years after the publication of Black's experiments, Henry Cavendish (1731-1810), with an apparatus designed after that of Hales, isolated "inflammable air", which we now call hydrogen, and he studied

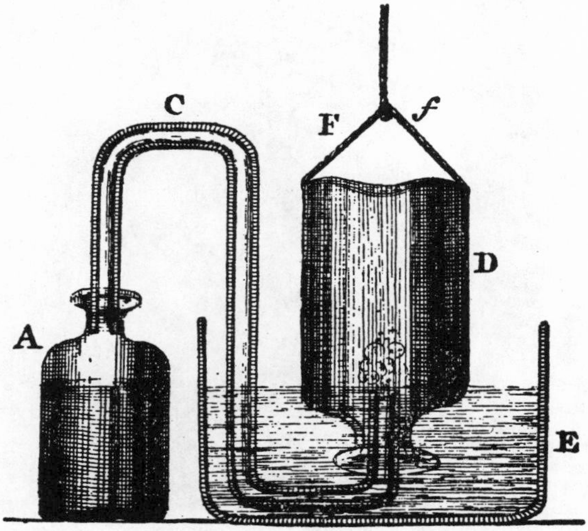

10. Cavendish's apparatus for the collection of gases (1766). A gas is evolved from the mixture in A and passes via the tube C into the water in D, into which it rises, displacing the water, which passes out into the trough E.

this gas and also Black's fixed air. He measured their densities and the solubility of fixed air in water. His apparatus (shown in Fig. 10) marks a further step forward in the collection of "airs." Two "airs" had thus been differentiated from common air, but the names given to them, "fixed air" and "inflammable air," show that it was still supposed that they were in some way related to the common air of the atmosphere and were mere modifications of it. Cavendish collected his "airs" over water. On one occasion he stored some fixed air over mercury because the "air" was soluble in water.

11. Priestley's method of collecting gases over mercury (1772–1774).

In 1772 the Rev. Joseph Priestley (1733-1804) took the further important step of collecting "airs" over mercury instead of water and was thus able to isolate a number of new "airs" that were soluble in water, the gases now known as hydrogen chloride, ammonia, and sulphur dioxide. Priestley's apparatus is shown in Fig. 11. Its invention, and especially the use of mercury in place of water, gave final form to the apparatus and technique required for the isolation of gases, as may be seen by comparing the apparatus of Boyle, Hales, and Cavendish, shown previously in Figs. 5, 9, and 10.

This in brief outline was the state of chemistry when Lavoisier was entering upon his first researches. The four-element theory was widely, if tacitly, accepted for want of a better one; Boyle's definition of the chemical element had fallen dead from his hands more than a century earlier; combustion was thought to be a process of decomposition, and so was calcination; the phlogiston theory dominated chemical philosophy; and the compositions of air and water were unknown and even unsuspected as were also those of that vast body of other less common and less familiar substances. Lavoisier changed all this; but such was chemistry when he entered that arena from which he was to depart one of the immortals of science.

V

EARLY RESEARCHES

Geology with Guettard

AFTER HIS YEARS AT THE MAZARIN COLLEGE AND AFTER HIS further studies under Lacaille, de Jussieu, Rouelle, and Guettard, Lavoisier was as well equipped for scientific work as any young man could be in the eighteenth century, for many branches of science were not then taught in the universities. He had shown where his interests lay, when in 1763, a year before he had completed his training at the Law School, he accepted an invitation to spend what time he had to spare in helping Guettard to collect details for a projected geological map of France. He spent three years on this work, mostly in the provinces bordering on Paris and in the districts of Brie, Vexin, Soissons, and Champagne, in surveying and in collecting specimens and, of course, continuing his observations with the barometer.

In the summer of each year he spent a recess at Villers-Cotterets with one or other of his great-aunts, Mme. Lavoisier and Mme. Prévost. Even on this annual holiday he continued his work, studying the geology of the forest of Villers-Cotterets, and going as far afield as Lagny, Chaumont, La Ferté-Milon, and Château-Thierry. At another time, he visited Beauce and went to Orléans and Moulins, and even to Thouars in Poitou, the farthest that he had yet been from Paris. From all these journeys he brought back numerous observations on the geology, mineralogy, and botany, of these places; and his notebooks show that he had his own views about his observations on the nature of the soil and of the fossil shells that he had collected. This was his first experience of original scientific study and also his first experience of travel: and for these three years he was assistant to Guettard, the leading geologist of France.

We have already seen a little of Guettard's character and personality; and here we might well refer, for further light on that intransigent and unrepentant individual, to the draft of a letter, declining an invitation from a lady, found among the papers that he bequeathed to Lavoisier. It reads:

> You should not be surprised that I am resolved to withdraw from society, but rather that I have been so bold as to remain there so long, having so little taste or talent for it. How can a man present himself in society who hasn't a fine hat, who has no fashionable wig, no embroidered coat, no jacket of the best dimity or breeches of like quality, no underwear of the finest linen? How dare he appear without stockings finer than a hair? Would he not be bold to present himself without neat shoes showing the instep and without buckles worthy of the trappings of Bucephalus? What word should one use to describe him if he has no shirt as fine as muslin and if it is not decked with cuffs and jabot of the finest Alençon or Flanders lace? Ah! Madame, what a man this is; he is a monster in society.

> Where does he come from? He is at least a Hottentot, a Tahitian. What can one do with this savage? Away with him, away with him, so that at least he shall not offend people or rather so that he shall stay in his den; he was not made to live with humankind. If only this badly brought up bear had some talent, but he hasn't the least; he doesn't play backgammon or chess; he knows nothing of piquet and even less of whist. What can one do with this animal? If he speaks, it is not of tassels and fittings and trimmings and crispings, of robes and coats and shirts and ribbons; he doesn't know the news, he doesn't read the newspaper. Away with him, away with him, the heretic! He is revolting! —This portrait is that of a man whom I know well; I recognize myself in it; and so, I have long since intended to withdraw from society, and I am now carrying out that intention to avoid embarrassing you with him. What could he do better? Nothing, I am sure; let him alone.

Guettard was a sincere Catholic, devoted to the cause of the Jesuits, under whom he had been educated; they had recently been expelled from France and their expulsion was a sore point with him. In disputes on the matter he was never sparing in offensiveness towards his opponents. "Few men," said Condorcet, "have had more quarrels."

First research: plaster of Paris

THE IMPORTANCE OF THESE YEARS IN LAVOISIER'S SCIENTIFIC development can scarcely be exaggerated. He was constantly at work with a man of rare genius, untiring industry, and independent spirit, collecting and classifying masses of information, observations, mineral specimens, and all facts bearing on the subject. Soon he produced his first research. It dealt with the different kinds of gypsum, the mineral from which plaster of Paris is prepared, and it had probably arisen in the course of the work upon which he was engaged with Guettard. The research was begun in 1764, when Lavoisier was twenty-one; it was submitted to the Academy of Sciences in 1765 and, owing to the usual delays, not published until 1768.

The memoir submitted to the Academy embodying the results of Lavoisier's researches on gypsum will always be of interest, since it was the first memoir of a series extending over thirty years and including some of the classic papers in the history of chemistry, and, indeed, in the history of science. In it Lavoisier determined the solubilities in water of the different varieties of gypsum. He explained the setting or binding of plaster by showing with the aid of the balance that gypsum lost a quantity of water when it was heated (forming what is called plaster of Paris), which it took up again when it was mixed afresh with water, and that this recombination or rather recrystallization was the cause of its setting or solidification. He noted that overburned or overheated gypsum would not bind with water to form plaster. He stated that he would not speculate as to the reason for this; he could make some suggestions or guesses, but speculation of that kind was out of place in the science of chemistry, where experiment must control every advance. The memoir was well written; the conclusions were clear and careful; theory was not allowed to pass beyond fact.

Lavoisier had taken the first step in his long career of scientific discovery.

Street lighting

SHORTLY AFTER HE HAD SUBMITTED HIS FIRST RESEARCH TO the Academy of Sciences, Lavoisier entered for a prize announced for competition by that body in 1765. Competitors were to present within a year an essay on the best means of lighting the streets of a large town at night. The special points to be considered were the degree of illumination and the easiness and economy of the means proposed; the prize was 2,000 livres. In preparing his essay, Lavoisier made a thorough study of the problem. He examined various kinds of lamps burning oil or candles, wicks and reflectors of different shapes, and the relation between the consumption of oil and the intensity of the illumination produced. In a way that was to characterize much of his later work, he made careful estimates of costs; and his figures reveal the dawning of those financial and economic powers that were subsequently displayed in an outstanding manner in the service of France. His undeviating purpose in the pursuit of knowledge led him to work, and to live, for six weeks in a dark room, which was hung with black curtains, and from which every trace of light was excluded, in order to make his eyes more sensitive to slight differences in intensity of illumination.

When the essays were examined, the Academy divided them into two classes. The first class contained those in which the writers had applied the principles of physics and mathematics to the problem, the second those that were purely practical. The prize was shared between three competitors whose essays fell into the second class. But among the competitors in the first class Lavoisier alone was named. His essay was specially mentioned and he was awarded a gold medal by the King. The medal was presented to him by the President of the Academy at the public assembly on April 9, 1766, some months before he was twenty-three; and the Academy resolved to publish the essay. The award was a very considerable distinction for a young scientist. The *Journal des Savants* reported that "this flattering distinction for so young an author, of which there is no previous instance in the Academy, has greatly pleased the public."

In the same month, April, 1766, Lavoisier resumed his geological observations, which had been interrupted by the preparation of the essay on street lighting; and he was soon geologizing in Corbeil, Arpajon, Reuil, and elsewhere near Paris, and, later in the year, he went again to Brie. His time was never unoccupied, as might be expected of a student and friend of Guettard.

There had, however, been a change in the household in the Rue du Four-St.-Eustache. Early in 1766, Mme. Punctis had died. Lavoisier's father, to simplify the formalities of succession, "emancipated" his son, according to the term used in French law, so that he might inherit what had been left to him before he had reached the age of twenty-five, the age of majority in French law. This step seems to have been taken with another end in view. It had now become clear that the young Lavoisier was not going to follow the family tradition of legal practice, although he had qualified at the Law School, but that he was intending to devote himself to science; the legal action taken by his father would put him in possession of the fortune left to him by his mother, which would provide him with an income for his scientific pursuits.

A geological tour with Guettard

LAVOISIER NOW HAD NOT ONLY SCIENTIFIC TRAINING, BUT scientific experience as well, and also a private income. A great opportunity soon presented itself. Guettard at last received official recognition and patronage for his work, his plan for a geological atlas of France being adopted by the Government. He had suffered much disappointment, and had long realized that he could not hope to complete the atlas in his lifetime, as the project involved many long journeys and the collection of extensive data, a task too heavy for one man and one span of life. So, when he was despatched by the Government in 1767 to prepare the maps of Alsace and Lorraine, he invited Lavoisier to join him, not as assistant, but as collaborator. The invitation was gladly accepted.

On June 14, 1767, at three o'clock in the afternoon, Lavoisier set out with Guettard for the Vosges; the travellers were mounted on horses, the roads being too bad for carriages. Necessary instruments had been obtained—a barometer and three thermometers, a hydrometer to determine the density of the water at various places, and a box of chemical reagents for making tests. Just before he left, Lavoisier recorded the reading of the barometer in his home and made arrangements for daily records to be made during his absence. His father and his aunt, anxious about his welfare while in the mountains of the Vosges and in Switzerland, gave him much advice about taking care of himself. Lavoisier, however, although devoted to his home and relatives, was a young man of twenty-four with fifty louis in his purse and a good horse between his knees; he was a keen scientific observer, looking forward to seeing new places and thoroughly aware that he was going to collect a mass of interesting facts; he was escorted by his trusty servant Joseph; and he was accompanying one of the most eminent French scientists. Life must have seemed good to him at that moment, even if the party had to be well armed to deal with any robbers they might encounter on the way; but it was clouded by the prospect of separation from his family for more than three months. That very evening, when they had got as far as Brie-Comte-Robert, he wrote home to say that the travellers were happy and well, as he was invariably to do in subsequent letters, and that the horses were fit and well foddered. This was the first letter of many; and in the Rue du Four-St.-Eustache, as Mlle. Punctis said later, the postman was awaited as if he were the Messiah.

It was not long, however, before Lavoisier yearned for his home, from which he seemed, after eight days, to have been absent for months. His thoughts turned often to his father and his aunt; and he had to remind himself that the journey was pleasant and that there was good work on hand, but he had to muster his resolution to overcome the feeling of home-sickness. A fortnight later, on June 28, the party reached Bourbonne-les-Bains, where both Guettard and the Lavoisier family had friends. While there was momentarily less anxiety in Paris, fears grew as the travellers reached parts that were strange to them and as they began to make diversions into the mountains of the Vosges.

Mlle. Punctis feared everything, the effects of the heat, the danger to the travellers of the very weapons that they carried for their protection, their projected visits to the mines; in expressing her fears to her nephew, she reminded him of his promise to take every care of himself. Then came another worry and a suggested solution: "We are afraid that you do not get all the letters that we write to you and your father suggests, if you think it suitable, that, in order that they should be given more attention in the post, they should be addressed *To Monsieur Lavoisier, on His Majesty's Service in the Vosges.*" Meanwhile, she kept Lavoisier posted in all the news of this bourgeois Parisian household, of the cat and the kitten, of the masons who were repairing the stables, of the card games that they continued to play without interest because neither Lavoisier nor "dear Doctor Guettard" was there to enliven them. With his father, she consoled herself for this enforced absence by arranging Lavoisier's library, adding to it the many books that he sent home, and unpacking the cases of mineral specimens that often arrived.

But Mlle. Punctis was deeply worried about the visit to the mines at Sainte-Marie and wrote: "Fear about the mines at Sainte-Marie obsesses me; what a relief for me when you will be on your way back from those mines, which cause all my worries!"

In another letter, she wrote: "In sending us your news, you show your affection for us; we need it more than ever. The further you go, the more my anxiety grows, especially as you draw near the mines at Sainte-Marie. Take care of yourself for a father and an aunt who live only for you. A letter scarcely reaches us but we are already waiting for the next."

Lavoisier's father wrote: "Send news as often as you can, a word in your hand that you are well with the date and place from which you write. We do not wish for more. You know how dear you are to us and therefore how anxious we become when several days pass without news from you."

Lavoisier could do no more than his best to allay the fears of his relatives. He was often far from the highroad and in places where there was no post or only one each week. Sometimes, he had to send his letters to the nearest town; and Paris was a month by post from Strasbourg and in some parts the four-horse mail coach took an hour to cover a league on the bad roads. Sometimes supplies had to be sent from Paris—more weights for the hydrometer, and new thermometers to replace those broken in use. Already it was suggested by Lavoisier that on the return journey his father should meet the party at Bourbonne-les-Bains and bring with him a new coat of green or grey cloth with a little gold braid. Mlle. Punctis consulted the tailor and found that green was more fashionable than grey and ordered a green coat of the best quality. There was, also, another request—that Lavoisier's father should bring a bowl of goldfish, which he should get from Marianne, Guettard's housekeeper at the Palais-Royal, where Guettard acted as curator of the collection of the Duke of Orléans. These were intended as a gift for Mme. de Brioncourt, whose hospitality the travellers had enjoyed on the outward journey. Lavoisier's father was a little taken aback at this request, but could not refuse, although he protested: "I shall have to hold the bowl of goldfish in my hands all the way and without any certainty of getting them there alive. This is an unpleasant task and very awkward for a traveller. I shall, however, forget the awkwardness and the difficulty by thinking that I am going to meet you again and embrace you."

Lavoisier's diary of this journey enables us to follow him day by day, even hour by hour, and shows the regularity of his work. Between five and six o'clock every morning, before starting out for the day, he recorded the readings of the thermometer and barometer; he repeated these observations at intervals during the day and concluded with a final one in the evening. Soil, relief of land, vegetation, were noted; quarries and works were visited and details of these visits were entered in the diary. The temperatures and densities of river and mineral waters were determined and recorded, and Lavoisier even went so far as to extend these observations to the water supplied at the inns at which the party stayed. Specimens of minerals were collected, both for an official and a personal collection. And so the journey wore on, sometimes comfortably and sometimes not, as, for instance, in one village, where the only accommodation was a granary, where a stock of onions was drying, which, as Lavoisier wrote in one of his letters, "tainted" the travellers.

The route lay through Provins, Troyes, Chaumont, Langres, Bourbonne-les-Bains, Vesoul, Villersexel, Lure, Ronchamps, where they visited a coal mine, and Luxeuil, to Giromagny, where they stayed for some days to go to Bussang and climb the Ballon d'Alsace (about 3,800 feet high). Lavoisier was much interested in the mountains of the Vosges and he wrote to say: "I have never seen anything in Nature that has impressed me so much." The great heat had now given way to storms and almost continual rain. And so on to Belfort, Montbéliard, and Altkirch, and then to the city of Basle, where they arrived on July 25, six weeks after leaving Paris. There they met a number of eminent scientists, including the mathematician, Daniel Bernoulli, Professor in the University of Basle. Lavoisier made a point of going to Domant to visit the tomb of the mathematician Maupertuis. His letters mention all these incidents and his impressions of Basle and the beauty of the Rhine, which he described as flowing there with great swiftness. An accident occurred here; the expedition's barometer was broken and no instrument-maker was available to repair it. Happily, their host, M. Jacques Bavière at the sign of "The Three Red Tankards," with the traditional hospitality of his kind, insisted on presenting his own barometer in its place, thereby unwittingly saving himself from utter biographical extinction.

After Basle, Guettard and Lavoisier visited Mulhouse, Thann, and Gérardmer, where they climbed to the highest point of the Vosges, the Grand Ballon (4,429 feet), accompanied by an artist whom they had engaged to sketch a panorama of the mountains for them and whom they kept with them as far as Colmar to make a series of drawings. On their arrival at Sainte-Marie, they found that the mines, which had caused so much dread to Mlle. Punctis, were flooded and that descent was impossible.

At Strasbourg, they met a number of prominent scientists, including the two chemists, Spielmann and Ehrmann. Lavoisier had a special interest here in meeting Brakenof, one of his correspondents in the collection of barometrical observations. He also bought a large number of chemical books, mostly German and unknown in Paris, to the value of five hundred livres, writing to his father to say that he was much afraid that this addition to his library had been rather expensive.

The route for the return to Paris lay through Saverne, Phalsbourg, Baccarat, Remiremont, Plombières, Épinal, Luxeuil, Mirecourt, and Nancy. It was planned to reach Bourbonne-les-Bains, where Lavoisier's father would be waiting, about October 6. The latter had set out from Paris and at Chaumont he received news that his son would reach Bourbonne on October 7 about one o'clock in the afternoon or later. He hurried on, as he had little time to spare. At Langres, he allowed his friends to detain him only an hour; he must get to Bourbonne that evening. At Montigny, there were no horses; but he insisted that he could not stay overnight and horses were taken from a field and harnessed to the post chaise and driven off in torrential rain. They covered scarcely a league an hour; and night came on while they were still in the woods. The weary traveller tried to dispel the tedium of his journey by peering into the darkness to see if any wolves lurked among the trees. At last he reached Bourbonne and comfort in the house of his friend, M. Robert, but he was up at six o'clock next morning ready to meet his son, who arrived with Guettard in the early afternoon.

After affectionate greetings, father and son had much

to say to each other; but the day's work had to be done and there were specimens to be labelled and despatched to Paris. Next day, they set out for home, Guettard and Lavoisier for Paris, Lavoisier's father leaving them at Chaumont as he had to go to Villers-Cotterets. Guettard and Lavoisier reached Paris safely on the evening of October 19 to be welcomed with relief by Mlle. Punctis. It had been a long and fatiguing journey, lasting more than four months; at times it had been very hot, at others there had been heavy and drenching rain. As the crow flies, it is 250 miles from Paris to the furthest point of the tour and, including the distances traversed in the surveys of Alsace and Lorraine, the travellers may well have covered more than a thousand miles. Yet, before retiring that night, Lavoisier at half-past eleven o'clock faithfully recorded the reading of the barometer. After two days at home, he joined his father at Villers-Cotterets, whence they both returned to Paris about the middle of November.

Lavoisier immediately turned to the task of arranging the information collected on the tour, setting out his numerous analyses of water in a lengthy memoir which, however, was not published until after his death. Along with Guettard he was involved in the heavy labor of preparing from their joint observations and other data the geological map of France, sixteen of the proposed 230 sheets of which were already engraved in 1770. But the expenses proved too great and there were delays; and many other affairs in France were already moving slowly. Presently, the retirement of Guettard led to the appointment of Monnet, Inspector-General of Mines, to carry on the work. In 1780 Monnet published the atlas in incomplete form, a key-map and thirty-one sectional maps covering northeastern France, in the joint names of Guettard and himself, ignoring all Lavoisier's contributions except those for the first sixteen sheets, but using without permission the wealth of other information which Lavoisier had collected. Lavoisier was galled at this treatment and protested at what he very properly called Monnet's "impudence." Monnet became an obstinate enemy and attacked Lavoisier's new system of chemistry even after its author's death.

However, the geological expedition of 1767 had been most successful and Lavoisier's happy collaboration with Guettard added to the reputation that he was already making among the scientists of Paris.

VI

ACADEMICIAN AND FARMER-GENERAL

THE ROYAL ACADEMY OF SCIENCES HAD BEEN INSTITUTED IN 1666 by Colbert at the direction of Louis XIV, with a small number of members. They received salaries and subventions from the Government for their work, and gave advice and drew up reports on official questions put to them for that purpose.

The organization of the Academy had been modified in 1699 and again in 1716; although its numbers had been increased, its membership was still limited, and its constitution was almost feudal. Members were divided into different grades, each with different privileges. In the highest rank were twelve honorary members, who must be noblemen, and from amongst whom the president and vice-president must be chosen. Next came eighteen *pensionnaires* or salaried members; and then there were twelve associates and, finally, in the lowest grade twelve assistants. In addition, there were honorary associates, retired *pensionnaires* and associates, and foreign associates.

Membership in each of the three grades, *pensionnaires,* associates, and assistants, was evenly divided between six sciences, geometry, astronomy, physics, chemistry, botany, and anatomy. The twelve assistants had few rights beyond bare membership; they sat on benches behind the chairs of the associates, and were permitted to occupy chairs only when any were empty. Only the twelve honorary members and the eighteen *pensionnaires* could vote in the affairs of the Academy and, therefore, in elections. When an election was to take place, the two associates in the science in which the vacancy had occurred were, however, allowed to discuss with the three *pensionnaires* in that science the names to be included in the list of candidates, but they could not vote.

Lavoisier's election to the Academy

FOR SOME YEARS LAVOISIER HAD BEEN CONSIDERED FOR ELECTION to the Academy. He was well known to many of the members, some of whom had been his teachers and were now his friends. They had put his name on the list of candidates as early as 1766, when he was still a young man of twenty-three. During his absence on the geological tour with Guettard in Alsace and Lorraine in 1767, his friends in the Academy had told his father how well-disposed their fellow Academicians were towards him.

Early in 1768 a vacancy occurred among the chemists through the death of Théodore Baron, known for his studies on borax; and it seemed that Lavoisier had a good chance of being elected, except for the presence of a formidable candidate, Gabriel Jars. Jars was a metallurgist who had done much work for the Government. He had superintended the development of lead mines in Brittany and coal mines in Anjou, and had made official visits to the mines of Saxony, Austria, Carinthia, Bohemia, the Harz, Sweden, and Norway, and the factories of Holland and England, whence he had brought back information of processes not known in France for the manufacture of red lead. He was thirty-six years old; and he was supported by the famous Buffon, treasurer of the Academy, and more significantly by the Minister, de Saint-Florentin, who was an honorary member and anxious that the important work that Jars had done for the State should be recognized by the Academy.

As for Lavoisier, he was twenty-five; he had been awarded a gold medal by the King, an unprecedented distinction, for his essay on the lighting of towns at night; he had submitted four memoirs to the Academy and these had been accepted; and his work on the geological survey of France, in collaboration with Guettard, who had spoken very well of him, had added to his reputation. It was also thought that a young man, who was well-informed, intelligent, and active, and who had an income that freed him from the necessity of earning his living in another profession, would be very useful in the scientific life of the Academy.

The election took place on May 18, 1768. Lavoisier obtained the majority of the votes, but the Academy merely recommended and the final choice lay with the King. The Minister submitted the two names to the King and advised that Jars should be elected to Baron's place; not wishing to go against the opinion of the majority of the Academy, however, he recommended that Lavoisier should be given a provisional place with the understanding that there would be no further election when the next vacancy occurred. The King gave his assent; and Jars and Lavoisier were formally admitted as assistants on June 1, 1768. In the Minister's letter, conveying the King's decision Lavoisier was described as "a very distinguished subject of His Majesty." Election at such an early age was an outstanding honor.

Lavoisier's election brightened the convalescence of his father, then recovering from a serious illness, and Mlle. Punctis was overjoyed that her nephew at twenty-five had obtained a distinction that usually came in the fifties after a great deal of work.

Lavoisier immediately entered actively into the life of the Academy, which met twice weekly, on Wednesdays and Saturdays from 3 to 5 o'clock in the Old Louvre (Plate 12). But attendance at its sessions was not the only duty of the members. It had always been a part of the Academy's work to prepare official reports on many practical matters of interest to the Government, as well as on memoirs submitted for acceptance and publication in their annual volume of *Memoirs*. Lavoisier was at once called upon to take his share, in association with other members, in these duties, which had always been heavy, as the Academy was in a sense an official body with funds and salaries provided by the State. During his membership Lavoisier took part in the preparation of reports on the water supply of Paris, prisons, mesmerism, the adulteration of cider, the site of the public abattoirs, the newly-invented "aerostatic machines of Montgolfier" (balloons), bleaching, tables of specific gravity, hydrometers, the theory of colors, lamps, meteorites, smokeless grates, tapestry making, the engraving of coats-of-arms, paper, fossils, an invalid chair, a water-driven bellows, tartar, sulphur springs, the cultivation of cabbage and rape seed and the oils extracted thence, a tobacco grater, the working of coal mines, white soap, the decomposition of nitre, the manufacture of starch, expressed oils, the distillation of phosphorus, the storage of fresh water on

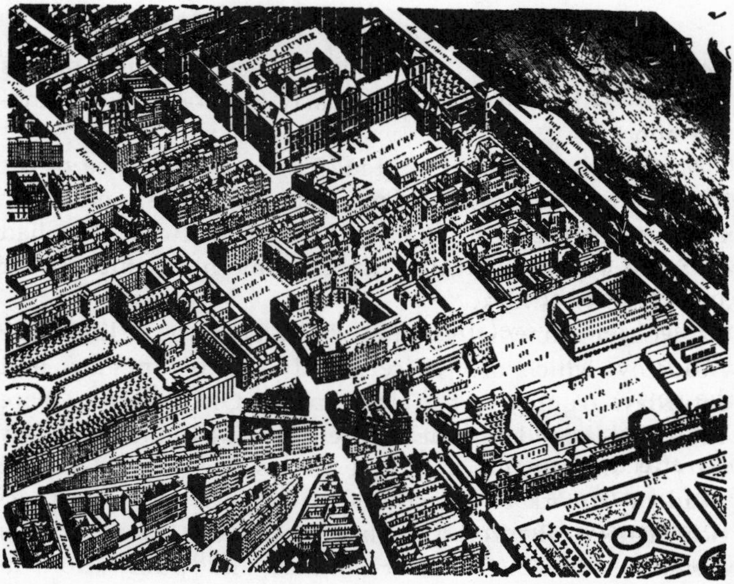

12. The Louvre and its environs in the eighteenth century. The Academy of Sciences met in the Old Louvre, still surrounded at that time by the new palace in course of construction, and no longer in use as a royal palace since the departure of the Court to Versailles. On the left, east of the Palais-Royal, is the Rue des Bons-Enfants, where Lavoisier lived after his marriage in 1771, until he moved to the Arsenal after his appointment as Commissioner of Gunpowder in 1775. The Hôtel de Longueville, headquarters of the Tax-Farm's tobacco monopoly, is shown in the right centre, the gate of its courtyard opening on to the Place du Carrousel. The Jardin de l'Infante can be seen between the Louvre and the river. *Musée du Louvre, Plan de Turgot.*

ships, fixed air, a reported occurrence of oil in spring water, lead ore, zeolite, marble, various machines, dyeing, the removal of oil and grease from silks and woollens, the preparation of nitrous ether by distillation, ethers, a reverberatory hearth, a new ink and inkpot to which it was only necessary to add water in order to maintain the supply of ink (some hopeful inventor seems to have sought the approval of the Academy), thermal springs, steel, vegetable rouge, specular iron ore, spathic iron ore, crystallization, the purification of oil, the art of the pewterer, salt, volatile marine acid, the decomposition of sal ammoniac, Chinese ink, glacial vitriolic acid, glass, the effects of mephitic fluids on animals, the decomposition of nitrous (nitric) acid, metallic soaps, a factory for watch glasses, the production of volatile alkali, the precipitates of iron, the estimation of alkali in mineral waters, a powder magazine for the Paris Arsenal, the mineralogy of the Pyrenees, wheat and flour, cesspools and the air arising from them, the alleged occurrence of gold in the ashes of plants, arsenic acid, the parting of gold and silver, the base of Epsom salt, the winding of silk, the solution of tin used in dyeing, volcanoes, putrefaction, fire-extinguishing liquids, alloys, the rusting of iron, a proposal to use "inflammable air" in a public firework display (this at the request of the police), coal measures, dephlogisticated marine acid,. lamp wicks, the natural history of Corsica, the mephitis of the Paris wells, the alleged solution of gold in nitric acid, the hygrometric properties of soda, the iron and salt works of the Pyrenees, argentiferous lead mines, a new kind of barrel, the manufacture of plate glass, fuels, the conversion of peat into charcoal, the construction of corn mills, the manufacture of sugar, the extraordinary effects of a thunder bolt, the retting of flax, the mineral deposits of France, plated cooking vessels, the formation of water, the coinage, barometers, the respiration of insects, the nutrition of vegetables, the proportion of the components in chemical compounds, vegetation, and many other subjects, far too many to be described here, even in the briefest terms. We shall refer to some of them in detail later. Work of this kind was congenial to him and matched his outlook in science; for he combined the vision and genius of the theoretical scientist with the practical mind of the technician and, as we shall see, of the administrator.

Jars died suddenly early in 1769 and the position in the Academy was regularized. No election was held and Lavoisier became one of the two assistants in chemistry.

Farmer-General

A FEW DAYS AFTER HIS NOMINATION FOR ELECTION TO THE Academy of Sciences, March 1768, Lavoisier took the most fateful, and indeed ill-fated, step in his life. He bought, after consultation with friends, a one-third share of the place of the Farmer-General Baudon in the Tax Farm and became Baudon's assistant.

When Lavoisier entered the Farm in 1768 under the lease (1768-74) of Julien Alaterre, there were sixty Farmers-General. The annual advance was fixed at 93,600,000 livres (francs), which amounted to 1,560,000 livres for each Farmer-General. A gratuity of 300,000 livres in addition had to be provided for the Controller-General. A Farmer-General was allowed an annual sum of 24,000 livres, office expenses of 4,200 livres, and an interest of 10 per cent on a million livres of his advance and 6 per cent on the remainder. But this total was subject to a deduction of 10 per cent for amortization, giving a net sum of 145,620 livres, a sum, however, not yet received by the Farm and still to be realized out of the future profits of the lease. Moreover, most of the Farmers had borrowed money, at an interest of 6 per cent together with legal charges, for their shares of the annual advance which the Farm had to make to the Government. The sum named was thus further reduced to 52,020 livres, out of which they had to maintain their families, pay their secretaries and clerks, educate their children, and also to provide pensions and incomes for the Minister's nominees.

For administrative purposes the Farm was organized in committees. There was a committee for accounts, one for the salt-tax, another for tobacco, for personnel, for litigation, and so on. The Farmers-General, although they bought their places, were nominated formally by the King. Every Farmer-General was a member of several committees, with days and hours fixed for meetings which he must attend regularly. Actual direction of affairs was in the hands of these committees; decisions of the committees were carried out by "correspondents", who gave the necessary instructions to subordinates. Every year some of the Farmers-General were chosen to make tours of inspection to survey their business throughout France. Members of committees, correspondents, inspectors, were all nominated from among the Farmers-General by the Minister, who thus controlled every department of the Farm. There was no president or central authority; every committee was master within its own house. The Farm was a very large organization with several great offices in Paris, and employed nearly thirty thousand officials throughout France. Its headquarters were at the Hôtel des Fermes in the Rue de Grenelle to the east of the Palais-Royal. Offices for the Paris customs and tolls were at the Hôtel de Bretonvilliers, bought by the Farm in 1719, at the east end of the Ile St.-Louis; those for the tobacco monopoly were at the Hôtel de Longueville, bought in 1749, in the Rue St.-Thomas near the Louvre; and those for the *gabelle,* or salt-tax, were near the Châtelet.

Lavoisier entered the Farm to invest and to develop by his own work the fortune left to him by his mother, so that he could devote himself more fully to the pursuit of science which, he felt, would require a great deal of money. His scientific colleagues did not regard his association with the Farm very favorably, not because they were opposed to the system, but because they feared that the work might interfere with his duties in the Academy. Fontaine, the geometer, consoled them, however, with the comment, "The dinners that he will give us will be all the better." Lavoisier did not allow the work of the Farm to deflect him from science, although his tasks were heavy. In the year of his entry into the company, he went on a tour of inspection in Picardy; during the years 1769 and 1770 other tours occupied most of his time; and there were committees to attend, reports to be drafted, problems to be studied. Although his association with the Farm was ultimately to bring him to the guillotine, there is much evidence, as we shall see later, to show that he constantly thought of the conditions of the poorest of his countrymen and that he never prospered unjustly. The Farm in Lavoisier's time was approaching in its methods the system of taxation that we are familiar with today; and the Minister, Necker, could report to Louis XVI in 1781 that "the Farmers-General, distinguished by their education, are not the financiers of former times." In the grim hours of revolution in 1794, however, only the old oppressions were remembered; the recent reforms were forgotten and the Farmers-General paid with their blood for the iniquities of their financial forefathers.

VII

EXPERIMENT REFUTES AN ANCIENT THEORY

IN PARIS IN THE LATE 1760'S THE SUPPLY OF GOOD QUALITY drinking water for the city was becoming a serious problem and a subject of general discussion. Lavoisier, no doubt, heard it talked of in his home. When he had been a student at the Mazarin College, he had passed through the district where water was drawn from the River Seine for distribution on the backs of men and horses. His interest would naturally be aroused by the proposal of the engineer Deparcieux to obtain further supplies by using water from the River Yvette. He had long been interested in spring and mineral and river waters, and in determining accurately their different densities; during his geological surveys, he had had extensive experience in the use of hydrometers.

Deparcieux died in 1768. Shortly afterwards his plan was criticized by the Carmelite Father Félicien of St.-Norbert, whereupon Lavoisier defended the scheme before the Academy of Sciences in July, 1769. He addressed himself not only to members of the Academy, but also to the public and to the civil administration of Paris, and he published his defence in the *Mercure de France* for October, 1769. When the Academy called upon him to report on proposals for erecting a steam pump to raise and distribute water from the Seine, he submitted a long report in 1770 with a detailed analysis of the costs of such an installation. The city authorities decided to adopt Deparcieux's plan, but their coffers were empty. Work was actually begun in 1788 in a form somewhat modified from that originally proposed, but ultimately the outbreak of the Revolution led to its suspension.

Although the prominent part that Lavoisier had taken in the official and the public discussions of this problem revealed once again his recognition of the service that science could render to the community, it served to remind him that chemical analysis was not yet sufficiently advanced to determine the purity of water for drinking purposes. He realized that the only scientific method of comparing two samples of water was a physical one, namely, the ascertaining of their densities; and even this method was not satisfactory, if it were true that water was partly convertible into earth, because its density would then be expected to vary accordingly.

At the Academy of Sciences discussions on the water supply of Paris would inevitably refer to the persistent belief that water in evaporating was partly converted into earth. Many chemists still believed this, although some others doubted it. It had been accepted by the greatest of all men of science, Newton himself.

Lavoisier, in the face of conflicting opinion, decided to experiment. His results were reported to the Academy on May 10, 1769, in two memoirs entitled "On the Nature of Water and on the Experiments alleged to prove its Transmutability into Earth," read before the Academy on November 14, 1770, and, owing to the usual delays, not published until 1773. In the first of these he gave a historical account of the work done by those who had preceded him on this problem; in the second he described his own experiments and results.

In dealing with the work of his predecessors, Lavoisier explained that he was concerned, not with the older theories of the philosophers about the elements, but only with facts. "I wish to speak of facts," he wrote. This was his invariable approach. Experiments alleged to prove that water was transmutable into earth fell into two groups in Lavoisier's summary. In the first of these, plants and vegetables were said to have been grown from water only; in the second, earth was said to have been produced by repeated distillation of water in glass vessels.

The classical experiment in the first group was that made by van Helmont on the growth of a willow tree, and described by him in 1648. We quote from his English translator:

> For I took an Earthen Vessel, in which I put 200 pounds of Earth that had been dried in a Furnace, which I moystened with Rain-water, and I implanted therein the Trunk or Stem of a Willow Tree, weighing five pounds; and at length, five years being finished, the Tree sprung from thence, did weigh 169 pounds, and about three ounces: But I moystened the Earthen Vessel with Rain-water, or distilled water (alwayes when there was need) and it was large, and implanted into the Earth, and least the dust that flew about should be co-mingled with the Earth, I covered the lip or mouth of the Vessel, with an Iron-plate covered with Tin, and easily passable with many holes. I computed not the weight of the leaves that fell off in the four Autumnes. At length, I again dried the Earth

> of the Vessel, and there were found the same 200 pounds, wanting about two ounces. Therefore 164 pounds of Wood, Barks, and Roots, arose out of Water onely.

Boyle had made similar experiments and stated in 1661 in his *Sceptical Chymist* that he had produced "mint and pompions" and "divers other vegetables of very differing natures out of water."

These observations, said Lavoisier, suggested that plants owed nothing of their growth or their parts to the earth in which they grew; and, since they left an ash or earth on being burned, that they were able to transmute water into earth. This conclusion, however, could not be justified without the most rigorous experiments, because it was known that two substances could combine to give a third substance totally different in properties from either of those from which it had been produced. Besides, in this case, it was not only the transmutation of water into earth that was alleged, but also its conversion into several plant constituents, including oils and resins and so on, which were not earthy substances.

On further reviewing these experiments, Lavoisier pointed out that it was important to consider the kind of water that had been used. Some of the experimenters who had repeated them had used common water, others river water or spring water, all of which contained earth and salts. Van Helmont had used rain water or distilled water, and rain water was known to contain salt, though not enough to account for the increase of 164 pounds in the weight of the willow tree.

Moreover, Hales had found that plants took in air through their leaves and that vegetables contained much air. Air, it was known, contained various volatile substances. So, while plants might appear to grow from water only, it must not be forgotten that they might take in the earthy matter already contained in water from certain sources, and also that they might derive some of their growth from the air and the substances contained in the air. Experiments on the growth of plants and vegetables in water did not, therefore, prove that water was convertible into earth; the earth might be already present in the water or derived from the air.

As for the experiments in the second group, in which earth was obtained by the distillation of water in glass vessels, some investigators had claimed that earth was continually produced from the same sample of water by repeated distillation. Others had explained this by asserting that the earth had come from the particles of dust always present in the air; and still others had pointed out that the earth was produced even when the water was distilled in a closed apparatus out of contact with air or even when water was violently agitated in closed vessels. Again, therefore, the experiments of the second group were indecisive and the source of the earth obtained in the evaporation or distillation of water remained uncertain.

In the second memoir Lavoisier described his own experiments. He decided to use rain water as the least impure source available. After rain had been falling for some time and had therefore cleansed the air of the foreign matter always floating in it in the form of dust, he collected a large amount of rain water in vessels, either of glass or of glazed earthenware, set on an open space away from trees and buildings. He evaporated a sample and found that it contained small amounts of a light earth and of sea salt. He then took some of this water and redistilled it eight times.

It was important to devise an experiment that would be

decisive. To Lavoisier the best method seemed to be to distil a weighed amount of water in a sealed vessel and to weigh the water and the vessel before and after the distillation. If the total weight of the water and the vessel proved to be greater after the distillation, it would suggest, as in Boyle's experiments on calcination, that fire had passed through the walls of the glass vessel and possibly combined with some of the water to produce earth. If the total weight remained unchanged, it would follow that the earth was produced either from the water or from the vessel in which it had been heated; but, if the earth had been produced from the water or from the vessel, there would be a corresponding decrease in the weight either of the water or of the vessel.

A suitable closed apparatus for continuous distillation and redistillation would be difficult to construct and likely to break. However, Lavoisier recalled the old apparatus used by the alchemists for such a purpose, in which a liquid was continuously distilled upon itself in a vessel picturesquely called a "Pelican." (Fig. 13)

A glass "Pelican" was therefore made of convenient size and fitted securely at the top with a glass stopper, well ground so that the stopper could be properly sealed with thick lute. In the experiment, the luted stopper was also wrapped with wet bladder and tightly bound with stout thread. A sensitive balance was obtained, the most accurate available, from one of the officials at the Paris Mint; the balance turned with less than one grain when loaded with five to six pounds.

The pelican was weighed after being washed with distilled water and dried, and a sample of water eight times distilled was put into it. It was then placed on a sand bath and heated gently, with the stopper in place but not sealed. At intervals the stopper was raised to allow some of the air expanded by the heat to escape and thus to avoid possible fracture of the vessel later on in the heating by the increase of pressure that would be produced. When the expansion of the air was considered to be over, the stopper was fully inserted, and the pelican was removed from the sand bath and weighed again. This

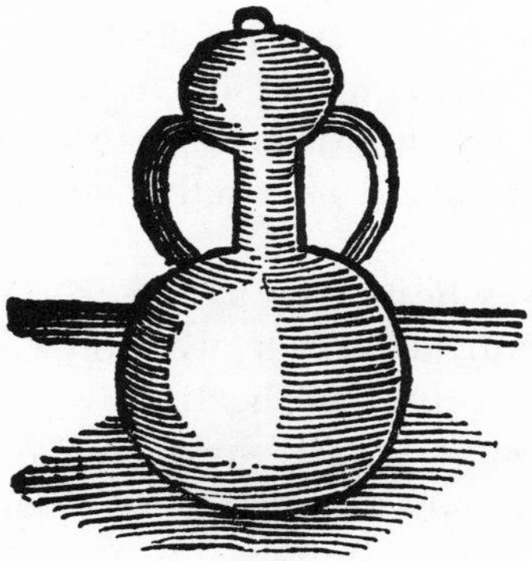

13. "The form of a Pellican," from John French, *The Art of Distillation* (London, 1651).

done, the stopper was luted and bound as already described; and the pelican was now set on the sand bath in such a way that the level of the water inside it was above the sand and visible changes could be easily detected.

It was October 24, 1768. The apparatus was ready and the method seemed adequate; the experiment would be long and tedious; the risk of fracture was great. The method was clearly inspired by Black's researches; all changes were to be followed by means of the balance. Implicit in this method, but not asserted, was a belief that matter was conserved in chemical change. The result might prove that Boyle was correct about his material particles of fire. On the other hand, there might prove to be no change in the total weight, but there might be a decrease in the weight of the water, in which case ancient theory would be upheld. Similarly, there might be a decrease in the weight of the vessel, and theory accepted for two thousand years and still taught by the most eminent chemist in Europe, Black himself, would be refuted. The problem of the water supply of Paris had led this young Frenchman at the age of twenty-five where practical problems lead all men of genius, to fundamentals—in this case to the question of whether matter was transmutable or not.

The sand bath was heated by a six-wicked lamp burning olive oil and trimmed every twelve hours. The temperature of the water during the heating, which continued for 101 days, was kept a little below its boiling point. For a long time, almost a month, no change was observed in the water, but on November 20 Lavoisier saw some very fine particles moving rapidly in it, when he had almost given up any hope of success, and even then the particles were so fine that he had to use a lens to see them. They appeared to be thin earthy flakes, and for about three weeks they seemed to increase in size. After December 15 there appeared to be no increase in either their size or their number, and presently they subsided to the bottom of the vessel, until at the end of February none could be seen moving in the water.

The experiment seemed to be complete. There was still much risk of an accident to an apparatus so fragile, and, as there appeared to be enough earth already produced in the vessel to decide the matter, Lavoisier resolved to stop at this point. The lamp was put out on February 1, 1769, and the apparatus allowed to cool. The binding and the lute were carefully removed from the stopper, an operation that must have caused him some anxiety, and the pelican was weighed without opening the stopper. There was no change in its weight, at least, the difference amounted to only a quarter of a grain, which was within the limits of precision of the balance. Continuous distillation of the water for more than a hundred days had thus produced no change in the total weight of the apparatus and, therefore, no external matter, either particles of fire or anything else, had entered the vessel.

The next step was to take the stopper out. This must have been difficult and involved further serious risk to the apparatus, since the air inside was now under reduced pressure. On loosening the stopper and raising it, Lavoisier heard a whistling noise as the air entered from outside which satisfied him that the sealing of the stopper had been satisfactory and that his method had worked just as if the pelican had been hermetically sealed.

Earth had clearly been produced in quantity; it remained to be determined whether it had come from the water or from the glass of the vessel. Lavoisier carefully poured the water and the earth from the pelican into a glass vessel and then he dried and weighed the pelican. It had lost 17.4 grains* in weight. The earth had therefore come from the glass of the pelican. But the weight of the earth produced should be equal to this loss. On weighing the earth removed from the pelican, Lavoisier found that it weighed 4.9 grains; the water in which the earth had floated yielded on evaporation to dryness 15.5 grains of earth. The total amounted therefore to 20.4 grains, which was 3 grains more than the weight lost by the pelican. Lavoisier ascribed this difference to the further and

* The Paris pound, the *poids de marc* of Charlemagne, was divided into 16 *onces,* each *once* into 8 *gros,* and each *gros* into 72 *grains.*

similar action of the water on the glass vessels in which it had been standing and in which it had been evaporated after removal from the pelican. The earth had clearly been derived from the glass, not from the water.

Water, therefore, was not transmutable into earth. Earth produced in the evaporation and distillation of water in glass vessels came from the destructive action of the water on the glass. Lavoisier had refuted a theory held for twenty centuries and accepted by many of his contemporaries.

The research that we have summarized in this chapter demanded much time and attention. Lavoisier, however, had an enormous capacity for work. There were meetings of the Academy to attend and reports to prepare; and he was occupied with tours of inspection for the Farm. From July, 1769, to January, 1770, his duties took him to Châlons-sur-Marne, Charleville, Épernay, Soissons, Lille, and Rheims. He began another tour in August, 1770, to Lille, Dunkerque, Gravelines, and Boulogne, and then to Rheims, Soissons, and Clermontois, his inspection occupying him until February, 1771. He was still working on the geological maps. With Lavoisier fundamental scientific research went hand in hand with practical affairs; the scientist and the public servant were indistinguishable.

VIII

MARRIAGE

IN THE FARM LAVOISIER'S CHARACTER AND ABILITIES HAD attracted the attention of the Farmer-General Jacques Paulze, a Parliamentary lawyer and a financier of the newer kind, skilful and upright, and one who had not hesitated to oppose the Minister, the Abbé Terray, when occasion demanded. Paulze had been director of the French East India Company and had considerable experience in public and state affairs. He had married in 1752 Claudine Thoynet, niece of the Abbé Terray, who then occupied the comparatively minor post of counsellor-clerk to the Parliament. Mme. Paulze died in 1761, leaving three children, two sons and a daughter, Marie Anne Pierrette, born on January 20, 1758, at Montbrison (Loire).

In 1771 Terray was Minister, Controller-General of Finance; Paulze was one of the ablest of the Farmers-General and reputed wealthy; Marie, his daughter and grand-niece of Terray, was thirteen years old and about to complete her education in the convent in which she had been placed after her mother's death. The Baroness de la Garde, who wielded considerable influence over Terray, suggested to him the arrangement of a marriage between her brother, the middle-aged and penniless Count d'Amerval, and Marie Paulze. Terray resolved to further the plan but he met with resistance from Paulze. Even though the father feared the reactions of the all-powerful Minister, his uncle, upon whom he depended for his place in the Farm, after a first tardy reply he wrote in terms which, as Grimaux said, did honor to his character:

> When you spoke to me, my dear uncle, of the marriage of my daughter, I looked upon it only as a very distant project and I thought, of course, that there would be some equality in age, character, fortune, and so on; I find, however, that such is not the case. M. d'Amerval is fifty, my daughter is only thirteen; he has not even 1,500 francs a year, and my daughter, although not wealthy, can at this moment bring twice as much to her husband; his character is not known to you, but that it cannot please either my daughter or you or myself, I am, indeed, well assured. My daughter has a decided objection to him; I will certainly not force her against her wishes.

Meanwhile, the Baroness de la Garde tried to overcome Marie's opposition by talking of her approaching departure from the convent and the brilliance of a presentation at Court. But all to no purpose; for Mlle. Paulze said that she disliked the Count intensely and that he was a "fool, stupid and ill-mannered, a kind of ogre."

Terray was greatly angered at Paulze's attitude and showed his resentment by threatening to remove him from his controlling place in the tobacco department of the Farm. Michel Bouret, whom we have already mentioned in a less worthy connection, protested vigorously in defence of his colleague, asserting that his energy and his talent showed that he was the only official competent to restore order in the different departments of the Farm, which were then in need of reorganization. Terray decided to leave Paulze in office, but he did not abandon the idea of the marriage.

Paulze, fearing further harassment from Terray, and wishing to save his daughter from further pursuit by d'Amerval, resolved that she should marry as soon as possible; he thought of his young colleague in the Farm. In November, 1771, a marriage was arranged between Mlle. Paulze and Lavoisier. The whole family was in perplexity as to what attitude Terray would adopt. His wishes had been disregarded; and his elder brother, Terray de Rozières, with whom he had quarrelled, had offered his house for the signature of the marriage contract, as Paulze's house was too small to entertain a large number of guests. The Abbé, however, accepted the decision with good grace, and not only promised to be present at the first ceremony, the signature of the contract, but also expressed a desire that the marriage should take place in his own chapel in the Rue Neuve des Petits-Champs.

At the signature of the contract, on December 4, 1771, two hundred guests were present, including statesmen, Farmers-General, scientists, and others. The Academy of Sciences was represented by the mathematician d'Alembert, the astronomer Cassini de Thury, and the botanist Bernard de Jussieu. Guettard was in Italy, but sent his congratulations from Rome. Lavoisier's devoted Aunt Constance and his great-aunt Mme. Lalaure were, of course, present; but in this distinguished company the notary in charge of the proceedings, possibly regarding them as somewhat insignificant, forgot to include them in the list of witnesses. Aunt Constance, however, having made her nephew heir to 50,000 livres at her death, claimed the privilege of signing the contract before all other witnesses, however highly placed; so did Mme. Lalaure, who had given her great-nephew a share in her estate. Paulze, not too well off at this time because his income during the first years as a Farmer-General on the current lease had brought him loss instead of profit, gave his daughter a dowry of 80,000 livres, a quarter of which was now passed to her, while the remainder was to follow during the next six years.

Lavoisier could not be charged with having married for money, for he was much wealthier than Mlle. Paulze. From his mother, he had more than 170,000 livres; his father gave him 250,000 livres in advance of his inheritance. He had borrowed nearly a million livres for the advance to be made by the Farm to the Government; and he had now a half-share in a place in the company, which meant that he had advanced 780,000 livres, while interest on the borrowed money reduced his annual income from the Farm to 20,000 livres.

The marriage was celebrated on December 16, 1771, in the private chapel, in the Rue Neuve des Petits-Champs, of the Controller-General of Finance, the Abbé Terray; he, with Terray de Rozières, his brother, acted as witnesses at the ceremony. Lavoisier was twenty-eight and his bride was nearly fourteen. They went to live in the Rue Neuve des Bon-Enfants, in a house which had been bought by Lavoisier's father at the corner of the Rue des Bons-Enfants and the Rue Neuve and which was demolished in 1864 to extend the Banque de France. Here Lavoisier lived for four years until his appointment to the Gunpowder-Commission in 1775 when he removed to the Arsenal.

The marriage was a happy one. Mme. Lavoisier was possessed of a high intelligence; she took a great interest in her husband's scientific work and rapidly equipped herself to share in his labors. Later she helped him in the laboratory and drew sketches of his experiments, two of which have survived and are reproduced here in Plates 26 and 27. She made many of the entries in his laboratory notebooks, all of which still exist. She learned English and translated a number of scientific memoirs into French; and her English translation of Richard Kirwan's *Essay on Phlogiston,* with comments by Lavoisier and others, was published and is one of the important eighteenth-century works on chemistry. She drew the diagrams for Lavoisier's classic treatise, *The Elements of Chemistry,* published in 1789. She learned to paint under the direction of David. The portrait of Lavoisier and his wife, reproduced here as the frontispiece, was, it is interesting to note, painted by David in 1788; it was until very recently regarded as the only authentic portrait of Lavoisier. Mme. Lavoisier painted a portrait, now lost, of Benjamin Franklin and it was greatly treasured by its subject. Writing from Philadelphia on October 23, 1788, Franklin said:

> I have a long time been disabled from writing to my dear Friend, by a severe Fit of the Gout, or I should sooner have return'd my Thanks for her very kind Present of the Portrait, which she has herself done me the honour to make of me. It is allow'd by those who have seen it to have great merit as a Picture in every Respect; but what particularly endears it to me, is the Hand that drew it. Our English Enemies when

> they were in Possession of this City and of my House, made a Prisoner of my Portrait, and carried it off with them, leaving that of its Companion, my Wife, by itself, a kind of Widow. You have replaced the Husband; and the Lady seems to smile, as well pleased.

In every sense, Mme. Lavoisier carried enough guns for her husband. She proved a charming hostess in entertaining his many friends and distinguished foreign visitors. Unfortunately, the marriage was childless and with Lavoisier's death in 1794 the family died out. Of Mme. Lavoisier's brave and devoted efforts to save her husband from injustice and execution, we shall speak later.

Shortly after the marriage, Lavoisier's father, as was the custom with rising bourgeois families, bought one of those many nominal offices that would confer the privilege of hereditary nobility on his son. The spirit of France was, however, changing; and Lavoisier made little use of his position as a member of the lowest rank of nobles. His name appears in the list of members of the "1789 Club" as "de Lavoisier." In the assemblies that were organized in the Revolution, he sat as a commoner.

IX

AIR AND FIRE

EARLY IN 1772 A MOST IMPORTANT DEVELOPMENT IN LAVOISIER'S chemical interests became apparent. He began to study combustion, and in almost fortuitous circumstances. Since the time of Boyle it had been known that strong heating would "destroy" diamonds; and now it had become a matter of dispute as to whether this "destruction" was an evaporation of the diamond into vapor or its actual combustion.

Is diamond combustible?

THE FIRST EXPERIMENTS ON THIS PROBLEM WERE MADE IN Cadet's laboratory by Pierre Joseph Macquer (1718-84), doyen of the French chemists and Director of the Royal Porcelain Factory at Sèvres, together with Cadet and Lavoisier. They heated some diamonds in a retort for three hours; the diamonds lost their polish and decreased in weight. Maillard, a jeweller, was so certain that diamond would not evaporate out of contact with air that he offered three diamonds for a further experiment, if he were allowed to be present and to arrange the apparatus to exclude air from contact with the diamonds. The experimenters accepted the offer and agreed to the conditions. Maillard put the diamonds in a clay pipe filled with powdered charcoal; he closed the pipe with lute and put it into a crucible, covered with chalk, then luted the whole arrangement inside two crucibles, one inverted over the other.

The experimenters were now free to apply a heat as great as they could command. But their furnaces had no effect; the diamonds were recovered with undiminished weight and luster, except for a slight darkening on the surface. The so-called destruction of diamond, therefore, needed air; but whether it was a process of combustion or evaporation was a matter for further research.

The next experiments were made by Lavoisier, Macquer, Cadet, and Brisson, in 1772 in the focus of the great burning lens of Tschirnhausen, owned by the Academy, in the noonday sun in the Jardin de l'Infante, the garden between the Louvre and the river. The experimenters had a room in the Louvre for their instruments and materials. Another lens by Tschirnhausen, of the same diameter but with a shorter focal length, was lent by the Comte de la Tour d'Auvergne, and de Trudaine offered the use of a much more powerful lens then under construction (shown in Plate 14).

It appeared from these new tests that diamond was a combustible. But of greater significance was the use of these powerful instruments, the great burning lenses. Smaller lenses had been in use long before; Boyle and Mayow and others had experimented with them. The large burning lenses were, however, not new instruments. In Germany towards the end of the seventeenth century Ehrenfried Walter, Graf von Tschirnhausen (1651-1708),

14. The great burning-lens constructed for the Academy of Sciences at the expense of Trudaine de Montigny. The large lens A was made by placing together two pieces of glass, each moulded as part of a sphere of diameter 16 feet; the hollow interior was filled with spirit of wine; the lens was a little over 4 feet in diameter and 6 inches thick at its centre. It was made at the Royal Factory at St.-Gobain under the direction of a committee of the Academy, consisting of de Montigny, Macquer, Brisson, Cadet, and Lavoisier. A movable lens B served for further concentration of the sun's rays, and the wheels I enabled the experimenters to rotate the apparatus horizontally on a stone emplacement about a vertical support C to face the sun. By means of the levers E, the lens A could be raised as required to a suitable height. The substance to be exposed to the action of the burning-lens was placed on an adjustable support G. The operator's eyes were protected by dark glasses.

Baron of Kieslingswalde and Stoltzenberg, who spent much of his wealth in scientific pursuits, built three glass factories and a polishing mill on his estates. He made large burning mirrors and four large burning lenses. The latter were three to four feet in diameter; and their manufacture was a considerable advance in technology. The Duke of Orléans presented one to the Academy and it was used from 1702 to 1709 to study the effects of heat on various substances. But, at that time, the main interest of the chemists lay in studying the changes produced by heat in the substances exposed to the action of the lens. The problem of what "air" might be given off by the substances had not yet arisen. Thereafter, the great lens of Tschirnhausen was put away in the Academy's collection of interesting apparatus; and there it remained for more than sixty years, until Lavoisier's expanding chemical interests led to its renewed use.

In the meantime, as we have seen, Hales had shown that air was contained in many substances. To obtain this air, he had heated various substances and collected the air evolved by means of the apparatus shown in Fig. 9 (p. 54). These results were, however, limited to the heat afforded by a charcoal or coal fire. Lavoisier was eager to apply a greater heat than this and thought of the Academy's large lens, which had been lying idle for two generations. Tschirnhausen's lens would enable the sun's rays to be concentrated in a small space; experimenters would have at their disposal a greater heat and a higher temperature. Lavoisier therefore suggested its use to the Academy on August 19, 1772, a suggestion that appears to have arisen from his appreciation of the importance of the work done by Hales, since he pointed this out to the Academy in making his proposals. It is clear that Lavoisier's thought was directed towards the air that could be removed from substances by subjecting them to fire and towards the interesting and now clearly established fact that diamonds, even when exposed to the most violent heat then attainable by science, were not "destroyed" unless they were in contact with air.

Phosphorus

LATE IN THE SAME YEAR, 1772, LAVOISIER BOUGHT SOME phosphorus, a substance so readily combustible that it is spontaneously inflammable in air and has, therefore, to be preserved under water. On September 10, he recorded in his laboratory notebook that he wished to find out whether phosphorus absorbed air when it burned. Hales had, in fact, already observed that burning phosphorus absorbed a considerable amount of air. Lavoisier did not complete the entry in his notebook and we do not know what results he obtained.

He must have made good progress with his experiments, however. On October 20, six weeks later, he sent a note to the Academy, reporting that burning phosphorus absorbed air in quantity; that it combined with this air to produce "acid spirit of phosphorus" (phosphoric acid), into the composition of which air, therefore, entered; and, further, that the phosphorus increased in weight by this combination with air. Altogether, this was a remarkable claim.

Twelve days later, on November 1, he deposited with the Secretary of the Academy a sealed note which was not to be opened until he so desired. The note read as follows:

> About eight days ago I discovered that sulphur, in burning, far from losing weight, on the contrary gains it; that is to say that from a pound of sulphur one can

obtain much more than a pound of vitriolic acid, making allowance for the humidity of the air; it is the same with phosphorus; this increase of weight arises from a prodigious quantity of air that is fixed during the combustion and combines with the vapors. This discovery, which I have established by experiments that I regard as decisive, has led me to think that what is observed in the combustion of sulphur and phosphorus may well take place in the case of all substances that gain in weight by combustion and calcination: and I am persuaded that the increase in weight of metallic calces is due to the same cause. Experiment has completely confirmed my conjectures: I have carried out the reduction of litharge [calx of lead] in closed vessels, with the apparatus of Hales, [heating the calx of lead with charcoal], and I observed that, just as the calx changed into metal, a large quantity of air was liberated and that this air formed a volume a thousand times greater than the quantity of litharge employed. This discovery appearing to me one of the most interesting of those that have been made since the time of Stahl, I felt that I ought to secure my right to it, by depositing this note in the hands of the Secretary of the Academy, to remain sealed until the time when I shall make my experiments known. Paris, November 1, 1772. Lavoisier.

In less than two months after embarking on his study of the combustion of phosphorus, Lavoisier was thus able to assert that phosphorus and sulphur on burning combined with air and gained in weight, the gain in weight being due to their combination with air; and he had a shrewd suspicion that the long-known gain in weight of metals on calcination was due to the same cause. It was air itself, not any part or constituent of the air, that he supposed had entered into combination with the sulphur and the phosphorus in his experiments, and that combined with the metals to form calces and was given off from them on reduction with charcoal. It was an astonishing conclusion: in another realm of human thought, it would have been treated as extreme heresy, because it utterly contradicted accepted belief. But it was based on facts and facts alone.

Towards the end of his life, when his theory had been extended and developed and when it was being accepted by his contemporaries, Lavoisier had cause to quote this note of November 1, 1772, in support of his claim that he had formulated his theory of combustion as early as 1772. It was the first step towards his new system of chemistry, although he was to modify it presently and to explain that it was only a part of the air that entered into these combinations. The sealed note was opened and read before the Academy on May 5, 1773, when further experiments had confirmed its conclusions.

Three months after he had deposited the sealed note with the Secretary of the Academy, Lavoisier wrote a long memorandum in his laboratory notebook on February 20, 1773, misdating it 1772. He had grasped the significance of the results of the experiments on the burning of phosphorus and sulphur and the reduction of calces; and he envisaged a long series of experiments on the air absorbed in combustion and set free from substances in every kind of chemical change. Hales and those who followed him, he noted, had considered that the air released from substances was identical with the common air of the atmosphere. Others, from Black onward, "observed differences so great between the air liberated from substances and that which we breathe, that they deemed it to be another substance, to which they have given the name of fixed air." But all the experiments so far made "come far short of the number necessary for a complete body of doctrine."

Further, went on Lavoisier: "It is established that fixed air shows properties very different from those of common air. Indeed it kills the animals that breathe it; whilst the other is necessary and essential to their preservation."

He continued:

> These differences will be exhibited to their full extent when I shall give the history of all that has been done on the air that is liberated from substances and that combines with them. The importance of the end in view prompted me to undertake all this work, which seemed to me destined to bring about a revolution in physics and in chemistry. I have felt bound to look upon all that has been done before me merely as suggestive: I have proposed to repeat it all with new safeguards, in order to link our knowledge of the air that goes into combination or is liberated from substances, with other acquired knowledge, and to form a theory. The results of the other authors whom I have named, considered from this point of view, appeared to me like separate pieces of a great chain; these authors have joined only some links of the chain. But an immense series of experiments remains to be made in order to lead to a continuous whole.

"An immense series of experiments. . . . destined to bring about a revolution in physics and in chemistry"—these are the words of inspired genius. The laboratory notebook in which they are enshrined is happily still in existence. It is one of the classic documents in the history of science.

With characteristic energy Lavoisier set to work on the researches that he had outlined. Less than a year later, in January, 1774, the first results were published in Paris in a volume entitled *Opuscules Physiques et Chymiques* or *Physical and Chemical Essays.* The work was divided into two parts, the first historical—an arrangement invariably chosen by Lavoisier, since he always saw the present evolving from the past, and the second describing his own experiments. In the first part he made a wide survey of all that had been already done on the air that was fixed in and separated from substances, and he pointed out that the chemists of France had taken no part in these important inquiries.

In the second or experimental part, Lavoisier described, first of all, a series of experiments which show that he had carefully repeated Black's work and had satisfied himself of the correctness of Black's explanation, which he adopted. He went somewhat further and supposed that it was Black's fixed air that combined with metals to form calces and increased their weight. However, he was more inclined to believe that it was common air, not fixed air, that entered into combination with the metals, or it might be an "elastic fluid," a gas, as we would now say, contained in the air. His reasons for this were that (1) metals could not be calcined in vessels exhausted of air; (2) the greater the surface of metal exposed to the air, the greater the calcination; (3) an elastic fluid was given off when calces were reduced with charcoal.

To confirm this last item of evidence, Lavoisier heated a mixture of minium (the red calx of lead, also known as red lead) and charcoal, both by means of Tschirnhausen's great burning lens and in retorts, and proved by collecting it over water, that an elastic fluid was given off in these reductions. (The surface of the water was covered with a layer of oil to prevent any of the "elastic fluid" from dissolving in the water.) In every case he obtained a considerable volume of the gaseous product, in one experiment a volume 747 times as large as the bulk of the residual lead.

Examination of this air showed that it was identical with Black's fixed air, obtained by the action of acids on chalk. Both proved fatal to animals that breathed them; both extinguished lighted candles and red-hot charcoal;

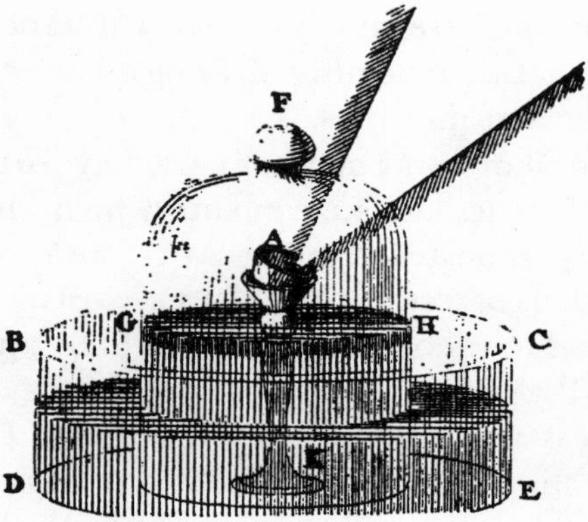

15. The combustion of phosphorous.

and both turned limewater milky. Therefore, the elastic fluid given off in the reduction of minium with charcoal was none other than fixed air, the carbonic acid gas or carbon dioxide of our day. This was a most important advance, possibly the most important in this series of experiments.

Lavoisier now turned again to the combustion of phosphorus, which had been the subject of his note of November 1, 1772, to the Academy. By means of a burning lens he ignited phosphorus in a known volume of air confined over water (Fig. 15) and found that the volume of the air was reduced between one-sixth and one-fifth. He repeated the experiment over mercury with a like result. The maximum reduction amounted to nearly one-fifth of the air. In a further experiment, Lavoisier estimated that 6 to 7 grains of phosphorus had gained in weight through absorption of air by 10 to 12 grains: and then, in a final test, he found that 154 grains of phosphorus had absorbed during its combustion 89 grains either of air itself or "of some other elastic fluid contained, in a certain proportion, in the air that we breathe."

The experiments were completed and the *Essays* published in less than twelve months. It was a heavy task for one occupied with other exacting duties. Lavoisier had not been able to conclude whether it was common air or fixed air or some other elastic fluid contained in the air that was absorbed in combustion and calcination. Despite his uncertainty, he had satisfied himself that these processes involved combination with some sort of air and concurrent gain in weight.

Lead and Tin

IN THE EARLY MONTHS OF 1774 LAVOISIER CONTINUED HIS experiments on calcination and on April 14 he submitted to the Academy a memoir describing the results. This memoir was read before the Academy on November 12 of that year and published in the December issue of Rozier's journal *(Observations sur la Physique)*. Here Lavoisier pointed out that his previous experiments had indicated that a part of the air or some matter contained in the air combined with the metals when they were calcined and increased their weight; that the release of air from metallic calces on reduction with charcoal confirmed this conclusion; and that he had proved that the air thus released was identical with Black's fixed air.

The calcination of metals had also been studied, as we

have seen, by Boyle, who had experimented in some cases in sealed glass vessels, and who had weighed the metal before calcination and then weighed the calx after removing it from the sealed vessel in which it had been produced. Boyle had found that the metals gained in weight and concluded that these gains in weight were due to material particles of fire which had passed through the pores of the glass and combined with the metals.

Boyle's method, said Lavoisier, was wrong. The sealed vessels should have been weighed before and after heating, and more importantly, before they were opened. If the gain in weight was due to combination of the metal with particles of fire which had passed through the glass, the total weight of the sealed and unopened vessel would have been found greater than it was before the calcination by the amount of the matter of fire that had penetrated it and combined with the metal.

On the other hand, continued Lavoisier, if the gain in weight of the metal was not due to the introduction of the matter of fire, but to combination with air or some other substance contained in the vessel, then the total weight of the sealed vessel after calcination would be the same as it was before. The vessel would merely contain less air; and it would be only at the moment of opening the vessel, when air entered from outside to replace the air that had combined with the metal, that an increase in the total weight of the apparatus would be detected.

In his experiments, therefore, Lavoisier, using weighed amounts of lead and of tin, repeated Boyle's work but with this very important change in procedure. The sealed vessels were found to weigh the same before and after calcination. Therefore the metals had not combined with particles of fire or any other external substance that had passed through the pores of the glass. On breaking the seals, the external air was heard to rush in with a whistling noise and the total weights of the vessels were found to have increased; and the increments were equal to the gains in weight detected in the metals when removed from the vessels and weighed separately. The metals on calcination had therefore combined with air contained in the sealed vessels.

This research led to some incidents that surprised Lavoisier. News of his experiments must have got abroad before he read his memoir to the Academy on November 12; for, on that day, Beccaria wrote to him from Turin to say that he had already shown, fifteen years earlier, that the extent to which tin was calcined in a sealed glass vessel depended on the volume of the vessel, that the total weight of the apparatus remained constant, and that there was a concurrent diminution of the air, and also that an account of this work had been given in the memoirs of the Turin Academy. Lavoisier arranged with Rozier, the editor of the *Observations,* in the December issue of which his memoir was to appear, that the relevant extract from Beccaria's letter should be published with his own memoir, and he accompanied this with a letter in which he stated that he wished to make it clear that he was unaware of Beccaria's experiments and that he had no desire to appropriate the discoveries of others.

He admitted that Beccaria had anticipated him in these experiments and added that he was, however, pleased to have such a confirmation of his work, although he had thought that he was the first to express such views about the calcination of metals. Here he was doing scant justice to himself; for his experiments were much superior to those of Beccaria and his conclusions had been carried further. But he was evidently disturbed about the matter. It may be that he was not given to reading original memoirs; but this seems improbable, since he had read widely when he was preparing his *Opuscules,* and had made extensive references in that book to the authors whom he had read. The same is true about the memoirs that he submitted to the Academy. It seems more probable that he had failed to notice the account of Beccaria's researches when he was studying the literature on this problem. His position was secure; for the theory that he was propounding on the gain in weight of metals on calcination was already contained in its germinal form in the work of Black and of Hales. No significance had so far been attached to it, however, and Lavoisier was taking it up with the intention of developing it alongside the study of the gases because it seemed to him "destined to bring about a revolution in physics and chemistry." While he was not the first to suggest that air played a part in calcination, he was the first to exploit the idea by means of quantitative experiments and to prove its correctness.

And there was more of this kind of thing to follow. Pierre Bayen, as we have seen, was much interested in the calx of mercury at the same time as Lavoisier. He had published several memoirs on its properties; and in one of these he had shown that mercury calx could be reduced to mercury merely by heating without the addition of charcoal—or, in contemporary terms, without the addition of phlogiston. In the course of his reading and apparently after the publication of his memoirs on this problem, he had found a copy of the *Essays* of Jean Rey (*c.*1582-*c.*1645), a physician of Le Bugue in Périgord. The book had been published in Bazas in 1630 and it presented an explanation of the gain in weight of lead and tin on calcination. The author supposed that the heat of the process of calcination made the air surrounding these metals denser by driving away its lighter parts, and that this condensed air adhered to the calx and increased its weight. It is to be noted that Rey did not believe that the calx was formed by combination of the metal with air. Evidently he thought that the calx was formed merely by the action of the fire, and that the condensed air adhered to the calx after the latter had been formed, not that the air combined with the metal to form the calx. In January, 1775, a letter from Bayen was published by the editor of the *Observations,* drawing attention to Rey's *Essays* and to the book's anticipation of the theory that was now being advanced to explain this well-known problem. He invited the editor to let the chemists of all countries know that "it was a Frenchman who by the power of his genius first divined the cause of the gain in weight shown by certain metals when exposed to the action of fire and converted into calces; and that this cause is precisely the same as that the truth of which has just been proved in the experiments described by M. Lavoisier at the last public assembly of the Academy of Sciences."

Rey's book was very rare. Today only seven copies are known to be extant; little mention of it seems to have been made between the time of its publication and Bayen's letter of January, 1775. Later, Lavoisier, while paying tribute to Rey's ideas, admitted that when he first saw the book he suspected from the modernity of its ideas that the date on the title page could not be correct and that the work might be a forgery. However, Rey's ideas were crude, even if interesting; and for the moment the essential improvement that Lavoisier had made in the theory seems to have escaped the notice of even Lavoisier himself. While others may have anticipated him (indeed, had) in the data of these experiments on the calcination of tin and lead and in crude theorizings that the increase in weight of these metals on calcination was due to air, he was the first to prove this to be the case, and the first to realize that this theory, applied alongside experiments on combustion and respiration, together with a study of the gases, was capable of bringing about a revolution in chemistry.

A month before Lavoisier read his memoir to the Academy, in October, 1774, Priestley had called on Lavoisier in Paris and had told him of the remarkable new "air" that he had obtained just before leaving England. This information had important consequences on the further development of Lavoisier's experiments; and we turn now to consider Priestley's discoveries made from August, 1774, to March, 1775.

X

PRIESTLEY'S "PURE AIR"

JOSEPH PRIESTLEY, ONE OF THE OUTSTANDING FIGURES OF THE eighteenth century, was distinguished in many fields. Born of sturdy Yorkshire and Puritan stock and nurtured in austerity, he was educated at the Dissenting Academy at Daventry. After some years as a minister, he was tutor at Warrington Academy from 1761 to 1767. Here he wrote his *History of Electricity*. He visted London every year and met Franklin, with whom he contracted a lifelong friendship. He was elected a Fellow of the Royal Society in 1766. From 1767 to 1773 he was minister of Mill Hill Chapel, Leeds, and from 1773 to 1780 librarian and literary companion to Lord Shelburne, later first Marquis of Lansdowne. In 1780 he became minister at Birmingham. His liberal and tolerant opinions on the French Revolution, the excesses of which he condemned, offended his political opponents and, with others who shared his views, he was the object of mob violence in the Birmingham riots of 1791. His chapel, his house, his library, and his apparatus, were destroyed; and, in danger of his life, he had to take refuge in flight to London. In 1794 after some continuance of less violent persecution during his residence in Hackney, he left for America where he died in 1804 at Northumberland, Pennsylvania. William Cobbett, who emigrated to America in 1792, concentrated the venom of Priestley's enemies into a scurrilous pamphlet published under the pseudonym of Peter Porcupine, but Priestley was defended from his native land in a generous tribute by the *London Monthly Review*, which asserted that the spirit that could dictate such outrageous abuse must disgrace any individual and any party.

Priestley gave early promise of being a scholar. He studied Greek, Latin, Hebrew, French, Italian, German, Arabic, Chaldee, and Syriac. He loved disputation and often chose the heterodox or the unpopular side in debate. He was, however, a kindly and courteous opponent, arguing for truth rather than victory. He wrote easily and fluently. The summation of his writings on education, theology, philosophy, and natural science, fills nearly twenty columns in the British Museum Catalogue; and *Observations on the Increase of Infidelity* stands cheek by jowl with *The Impregnation of Water with Fixed Air*. Even on his deathbed, he was correcting proof sheets.

While Priestley was librarian to Lord Shelburne, his duties were light and he had ample opportunity for continuing the chemical experiments that he had begun in 1768 when at Leeds. In the summer of 1774 he obtained a large burning lens, twelve inches in diameter and with a focal length of twenty inches. It was not so large as the great burning lenses of Tschirnhausen that we have already mentioned, which were used by Lavoisier and others in Paris, and which were thirty-three inches in diameter. But Priestley made a far more important advance with his lens.

He had planned a series of experiments in which various classes of substances were to be subjected to the great heat and high temperature of a burning lens in order to see what kinds of air they would yield. He was, like Lavoisier, continuing the work begun by Hales. He

16. The glass vessels in which oxygen was first isolated by Priestley in 1774.

took some small glass vessels, *a a a* (shown in Fig. 16), placed the substance to be heated in the bottom of one of them, filled it with mercury, and then inverted it in a basin of mercury. The burning lens was then gradually focused to heat the substance, which was now confined by the mercury against the top of the inverted vessel.

One of the first substances that Priestley heated in this apparatus was known as *mercurius calcinatus per se* or mercury calx. It was a red powder produced by heating mercury for a long time in air, and was, in the terms then used, calx of mercury. Turning the large burning lens into position in the bright sun at Bowood, the country house of his patron at Calne in Wiltshire, Priestley found that air was readily given off. He collected enough to carry out some tests. The air proved to be insoluble in water; a candle burned in it with a remarkably vigorous flame, and a piece of red-hot wood sparkled in it and was rapidly consumed.

This famous experiment was made on Monday, August

1, 1774. It has been stated that August 1 in that year fell on Sunday and Priestley has been pictured as spending his Sabbath in ways not suited to a rigid Dissenter. But the historic day was a Monday, as reference to a perpetual calendar will show.

For the moment, experiments ceased. Lord Shelburne, accompanied by Priestley, set out for the Continent later in the month of August. In October, the travellers were in Paris, and there Priestley was welcomed by the members of the Academy of Sciences. His scientific researches were well known and had brought him great reputation. Two years earlier, in 1772, he had been awarded the Copley Medal, the highest distinction in the gift of the Royal Society of London, for his researches on different kinds of "air" and on electricity. Among those who greeted him warmly in Paris was Lavoisier.

In the course of their conversations about chemistry Priestley told Lavoisier of the new air that he had obtained two months previously from mercury calx and he spoke of its properties. He returned to England in November and resumed his experiments. On March 1, 1775, he discovered that his new air was fit for respiration. This was a most unexpected result.

Priestley, who is one of the greatest of scientific experimenters, had already devised an excellent method of determining what he called the "goodness" of common air, that is, its fitness for respiration. He had found that nitrous air, which he was the first to isolate and study in 1772, reacted with common air to an extent that corresponded with the "goodness" of that air. His experimental test was made by adding one part of nitrous air to two parts of the common air to be tested, which was confined over water. We represent his method diagrammatically in Fig. 17 (a). After addition of the nitrous air, a contraction occurred, because the products of the chemical change that had occurred were soluble in the water, which therefore rose in the vessel. It rose to the mark 8 (Fig. 17 (b)), which corresponded to a contraction of two parts in ten, or one-fifth of the volume of air originally taken for the test. This was the greatest reduction obtained and corre-

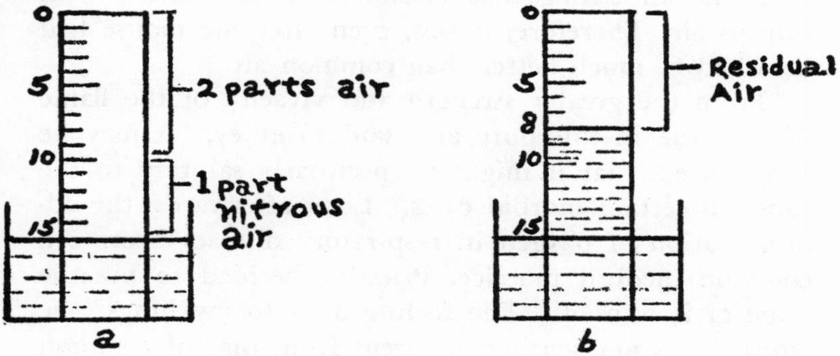

17. Priestley's test for the "goodness" of air.

sponded to the highest degree of "goodness" of any sample of common air. It is an excellent result: today we know that it corresponds to the normal proportion of oxygen in the atmosphere, namely, one-fifth.

Before he had invented this method, Priestley had used mice to test the "goodness," but he had found that they did not give satisfactory results. Some mice lived longer than others, even twice as long, in samples of the same air. With the nitrous air, he had a reliable and exact method at his command. What Priestley called nitrous air is now known as nitric oxide; it combines with the oxygen of the air and the resulting products dissolve in water.

When Priestley applied his test to the air obtained from

mercury calx, the air proved to be, as he first thought, at least as good as common air. Therefore, it must be fit for respiration. A week later, on March 8, 1775, Priestley put a mouse into a sample of the new air. The mouse breathed the air without any ill effects and survived for much longer than it would have in common air. Moreover, the air still gave a considerable contraction with nitrous air. Therefore, it was, even after the mouse had breathed it, much better than common air.

"From the greater strength and vivacity of the flame of a candle in this pure air" said Priestley, "it may be conjectured, that it might be peculiarly salutary to the lungs in certain morbid cases." Long afterwards, the administration of oxygen in respiratory diseases became a common medical practice. Priestley decided to breathe some of it himself. "The feeling of it to my lungs," he wrote, "was not sensibly different from that of common air; but I fancied that my breast felt peculiarly light and easy for some time afterwards. Who can tell but that, in time, this pure air may become a fashionable article in luxury. Hitherto only two mice and myself have had the privilege of breathing it." It is to be noted that Priestley described it as "pure air."

Rough estimates of the density of the new air, made by comparing the weights of samples of it and of other airs contained in a bladder, showed that it was a little heavier than common air and that the residual air left after combustion had taken place in common air was a little lighter than common air.

On March 15, 1775, a week after the first experiment with the mouse, Priestley reported his discovery in a letter addressed to Sir John Pringle, President of the Royal Society. The letter was read to the Society on March 23 and published in the *Philosophical Transac-*

tions. It was the first public announcement of the isolation of the new air and the discovery that it was respirable. These experiments were made in London at the town house of Priestley's patron, Shelburne House, later Lansdowne House, Mayfair.

But it was then commonly believed by chemists that in respiration phlogiston was exhaled from the lungs, just as it was given off by all burning bodies, and that in both processes it was absorbed by the common air of the atmosphere after its emission from the breathing animal or the burning body. Since it was known that both life and fire were extinguished if they were not continuously supplied with fresh air, it was supposed that a limited amount of air could absorb only a limited amount of phlogiston, as a sponge can absorb only a limited amount of water. And so a flame died out and an animal perished without fresh air, because the air about them had become saturated with phlogiston and could absorb no more of it. The fitness of any air for combustion or respiration depended, therefore, on its capacity to absorb the phlogiston evolved in combustion or exhaled in respiration. As the new air seemed to be many times better than common air for these processes, it must, Priestley thought, be free from phlogiston, and he therefore named it "dephlogisticated air." Stripped of its theoretical implications, this means, in effect, that Priestley thought that what he had obtained was pure air; and, indeed, he had so described it, as is seen in the passages quoted above and as was natural, since, in his view, the common air was always impure through its contamination with the phlogiston from fires and furnaces and respiration.

From his historic letter of March 15, 1775, historic because it contains the first account of the gas now known as oxygen, we quote:

At the time of my last publication I had not a large burning lens. . . . I have now procured one of twelve inches in diameter; and the use of it has more than answered my highest expectations. The manner in which I have used it, has been to throw the focus upon the several substances I wished to examine, . . . when confined by quicksilver, in vessels filled with that fluid, and standing with their mouths immersed in it. I presently found that different substances yield very different kinds of air by this treatment. . . . But the most remarkable of all the kinds of air that I have procured by this process is, one that is five or six times better than common air, for the purpose of respiration, inflammation, and, I believe, every other use of common atmospherical air. As I think I have sufficiently proved, that the fitness of air for respiration depends upon its capacity to receive the *phlogiston* exhaled from the lungs, this species may not improperly be called *dephlogisticated air*. This species of air I first produced from *mercurius calcinatus per se,* then from the red precipitate of mercury, and now from red lead. . . . That this air is of that exalted nature, I first found by means of nitrous air, which I constantly apply as a test of the fitness of any kind of air for respiration, and which I believe to be a most accurate and infallible test for that purpose. Applying this test, I found, to my great surprise, that a quantity of this air required about five times as much nitrous air to saturate it, as common air requires. . . . A candle burned in this air with an amazing strength of flame; and a bit of red hot wood crackled and burned with a prodigious rapidity, exhibiting an appearance something like that of iron glowing with a white heat, and throwing out sparks in all directions. But to complete the proof of the superior quality of this air, I introduced a mouse into it; and in a quantity in which, had it been in common air, it would have died in about a quarter of an hour, it lived, at two different times, a whole hour, and was taken out quite vigorous. . . .

The results of Priestley's researches provided a further clue to the closer solution of the problem to which Lavoisier was now directing his attention.

XI

MERCURY CALX: GUNPOWDER

IN OCTOBER, 1774, WHEN PRIESTLEY WAS IN PARIS, HE HAD, AS we have just seen, informed Lavoisier of the results of the experiments that he had made on mercury calx in August of that year. Earlier, in February, 1774, Pierre Bayen (1725-98), Master Apothecary to the French Army, had reported that calx of mercury, unlike other calces, could be converted into metallic mercury on heating without the addition of charcoal, and, moreover, that it lost in weight in the process. This property of mercury calx was therefore known to the French chemists before Priestley brought the further information that on heating it gave off an air in which a candle burned with a remarkably vigorous flame and a piece of red-hot wood "consumed very fast." Further, on November 19 of the same year, with Sage and Brisson, Lavoisier reported to the Academy that the statement made by Cadet, chief apothecary at the Invalides, on September 3, 1774, that mercury calx could be reduced without charcoal, had been found to be correct.

Metals combine with pure air

THERE WAS, THEREFORE, MUCH INTEREST AMONG THE PARIsian chemists in this substance, mercury calx. After Priestley had told of his far more important observation, Lavoisier proceeded at once, in November, 1774, to experiment on it; and further experiments followed in February and March of 1775. Then on April 26, 1775, he read before the Academy a memoir entitled "On the Nature of the Principle which combines with the Metals during their Calcination and increases their Weight." In it we see the first results of his conversation with Priestley, the full consequences of which were, however, not realized for some time.

In describing his work, Lavoisier began by saying that he could not in the limited time at his disposal enter into the question of whether there were different kinds of air or whether there were different modifications of atmospheric air, but that he would show that it was air itself, "undivided, without alteration, without decomposition," that combined with the metals on calcination.

Most metallic calces could not be reduced to metals unless charcoal were added; and in the reduction, the charcoal was consumed. It therefore followed that the air given off in the reduction was not a simple substance, but was in some way a combination of the elastic fluid separated from the metal and of that discharged from the charcoal. That the elastic fluid proved to be fixed air was no warrant for concluding that it existed as fixed air in the calx before reduction.

These circumstances, said Lavoisier, indicated that experiments should be directed to calces that were reducible without addition of charcoal. Calx of mercury seemed most suitable for the purpose. To satisfy himself that this substance was a true calx, Lavoisier first reduced it with charcoal and collected the elastic fluid given off. It proved to be fixed air; it dissolved in water, asphyxiated animals, extinguished fire and flame, gave a precipitate with limewater, and combined with caustic alkalis to give mild alkalis. It was unquestionably fixed air; and therefore mercury calx was a genuine calx.

In the next experiment, Lavoisier heated an ounce of calx of mercury in a small retort and collected the issuing air, in the now familiar way, over water in a receiver filled with water. He gave no diagram of this apparatus until many years later. The volume of air obtained was seventy-eight cubic inches. Its properties differed completely from those of fixed air. It did not dissolve in water; it gave no precipitate with limewater; it did not combine with alkalis or diminish their causticity; it could be used again to calcine metals; it was diminished by nitrous air (Priestley's test); it was not fatal to animals, but supported their respiration; and lighted candles and burning bodies, far from being extinguished in it, burned with enlarged and more brilliant flames. It appeared to be, said Lavoisier, not merely common air, but common air in a purer state, fitter for respiration and combustion. All this was much the same as Priestley had said, shorn, however, of the terms of the phlogiston theory.

Therefore, concluded Lavoisier, the principle that combined with metals during their calcination and produced an increase in their weight was the pure air of the atmosphere. The memoir announcing these results was published in May, 1775, in Rozier's *Observations*.

At the conclusion of his researches of 1773, published in the *Opuscules* (1774), Lavoisier had been uncertain whether it was Black's fixed air, or air itself, or some substance in the air that combined with the metals in calcination. He had seemed to favor the first of these possibilities. In November, 1774, he had concluded that it was air itself in its pure state. Now in April, 1775, after Priestley's results of August, 1774, were known to him, he became more convinced that the substance involved was the pure air of the atmosphere, which was recovered in its pure state from the calx of mercury when the calx was reduced without charcoal, whereas, if charcoal were used, the air given off in the reduction was fixed air.

The influence of Priestley's work on Lavoisier's researches is already clear, but its full effect followed later. Lavoisier's immediate hesitations and doubts were resolved. Common air in its pure state, not fixed air, would give the ultimate answer.

The Gunpowder Commission

AFFAIRS OF STATE NOW BROKE IN UPON LAVOISIER'S ALREADY busy life. Louis XVI succeeded to the throne of France in 1774. In 1775 the Liberal Turgot was called by the new King to replace the Abbé Terray as Controller-General of Finance. The new Minister could do little to reform the Farm immediately, except to compel the Abbé to disgorge the sum of a hundred thousand crowns, which he handed over to the hospitals of Paris, but he took steps for the future. Henceforth appointments to the Farm were to be restricted to those whose ability and knowledge would be useful to the State. Turgot then turned his attention to the national production of gunpowder.

The gunpowder required by the French army and navy was supplied by a company organized like the Tax Farm; and it was the story of the Tax Farm retold. The company of financiers who took over the powder supply contracted to provide a million pounds of gunpowder every year. If they failed to deliver this amount, they were not compelled to make it up later; nor had they to provide extra supplies in any national emergency, if, for instance, France was at war. Thus France had to buy foreign gunpowder at very high prices in time of war, which was one of the causes that brought about the end of the Seven Years' War in 1763.

There was nothing, either motive or interest, to urge or compel the company to increase the output of gunpowder. They were not allowed to pay more than six sous a pound, while the cost of manufacture was twelve sous a pound. The production of powder depended on the production of its ingredient saltpeter, and the maximum price payable for this was seven sous a pound, which was less than the cost of production. The Government had therefore to indemnify the manufacturers of saltpeter. The company had no permanence; their lease was limited to six years; and they were concerned with making money, not gunpowder. In the Seven Years' War, France bought saltpeter from the Dutch at prices up to twenty-five sous a pound and gunpowder at three francs a pound.

The saltpeter makers enjoyed privileges that were a public burden. They could claim free lodging and transport; they had the right to dig anywhere for saltpeter; and they could be bought off by those who did not want their attentions. From such an abuse, they reaped a profit while the nation went short of powder. Saltpeter was found in soil that had been much trodden by cattle and mixed with their excreta. Stables, sheepfolds, pigsties, and dovecotes, were the usual places of search. The material was dug out, dried, and mixed with lime and wood ashes; it was then leached with hot water, and the solution obtained gave saltpeter when evaporated. Hot water, wood ashes, and evaporation, required a supply of wood for heating and for ashes. The saltpeter makers were entitled to buy the necessary wood at a very low price. Free lodging, free transport, and cheap wood, had to be supplied by the parishes in which they worked; the cost to the nation was very high when all was taken into account. Yet the output of saltpeter, and therefore of gunpowder, was decreasing all the time. The company enjoyed a monopoly for the sale of saltpeter, blasting powder, sporting powder, and the low-grade powder sold to countries engaged in the slave trade; their annual profits were as high as 30 per cent. And France went short of gunpowder!

Turgot's reforming hand was clearly needed here. Lavoisier drew up a report on the system with estimates of the current cost to the nation. Although the company's lease had still four years to run, Turgot cancelled it and by a decree of March 30, 1775, established a Gunpowder Commission. Four Commissioners were nominated on June 30 to carry out all the duties previously falling on the company. One of these was Lavoisier, as well known for his knowledge of chemistry, so essential for this kind of work, it was stated, as for the diligence, ability, and integrity, that he had shown in the office of Farmer-General of Taxes.

The Treasury lacked the money to indemnify the company for the annulment of their lease and for their buildings, apparatus, and materials. The Commissioners were to advance four million livres to pay these charges at 1 per cent above legal interest over a period of four years; and this sum was to be repaid to them as soon as the reformed system allowed. For their services, the Commissioners would receive 2,400 livres, with additional allowances according to the amounts of powder and saltpeter produced. The aims of Turgot and Lavoisier were to free the nation from the levies of the saltpeter makers, to benefit the Treasury by the profits of the monopoly in the sale of powder, and to increase the national output of saltpeter. These ends were rapidly achieved.

The decree instituting the Commission relieved the villages from the burden of providing free transport, and gave them the right to bargain for the prices of lodging and of wood. Later, in 1778, the Commission introduced a further reform by demanding official action restricting the right to dig to stables, sheepfolds, and dovecotes.

Within ten months of the change, the profits amounted to nine hundred thousand livres, and in less than two years half the Commissioners' advance was repaid. The yield of saltpeter increased, despite the restrictions of search introduced at the instance of the Commissioners. Villages were no longer anxious to buy themselves off, and it was in the interest of the saltpeter makers, who were now better paid, to increase production as much as possible.

Still, supplies were inadequate for the national needs and the Commissioners were compelled to buy further supplies from abroad. To avoid this, from the moment of their entry upon their duties they had studied ways to increase the national output. In other countries artificial niter beds had been established, but in France little was known about this method, which was, however, quite simple. It reproduced the conditions in which saltpeter was formed in stables, sheepfolds, and dovecotes. Animal excreta, decaying vegetable matter, and either old mortar or earth containing chalk, were mixed in dry ditches which were then roofed to keep out the rain. The mixture was turned now and then, so that air might penetrate it, and a little water was added. After some months, the chemical changes were complete. Wood ashes were then added and the mass was leached with hot water, which dissolved the saltpeter that had been formed. The solution obtained was then evaporated and saltpeter was left.

But in France this manufacture had not been tried and research was therefore called for. On Lavoisier's advice, Turgot directed the Academy to announce that a prize would be awarded in 1778 for the best essay on the most advantageous method of making saltpeter, and to draw up a detailed program which set out concisely what was known about the formation of saltpeter, what books were to be consulted, and what experiments were to be tried. Without the customary interval of eight days, the Academy immediately nominated a committee for prompt action. Lavoisier was appointed as one of the members. Some objections were made when his name was mentioned, because he was one of the Gunpowder Commissioners, but it was pointed out that the Commission was no longer a private company. Within a month, a program was drawn up and three thousand copies were printed and distributed. All possible sources of information on the manufacture of saltpeter were explored, at home and abroad, and the information collected was published in 1776. Lavoisier in 1777 edited for the Gunpowder Commission a volume on the erection of niter beds and the manufacture of saltpeter.

Turgot's reforms and his liberal outlook had, however, been too much for the powerful Court circle. After thirteen months in office, he was dismissed in 1776. The work of the Gunpowder Commission continued, largely under Lavoisier's direction, and its institution proved to be one of the most valuable reforms effected by Turgot. By 1778 there was enough powder in the magazines for three campaigns.

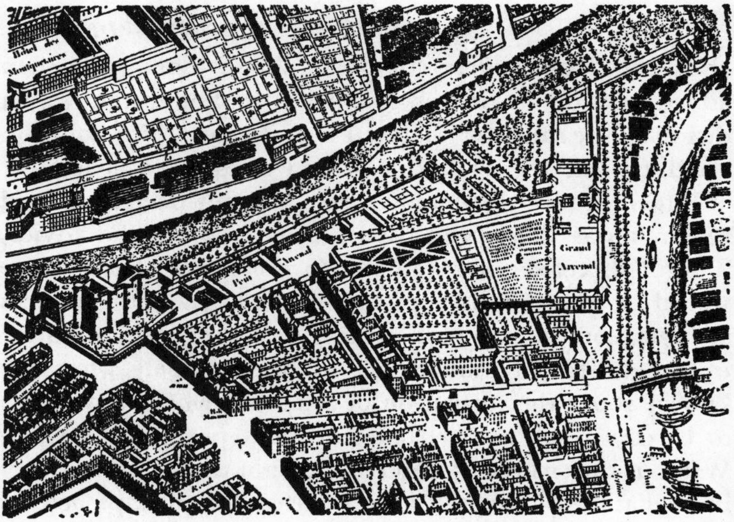

18. The Paris Arsenal in the eighteenth century, showing the Grand Arsenal and the Petit Arsenal with Port-St.-Paul, on the right, for cargoes to and from the Arsenal. The entrance to the Grand Arsenal was near the convent of the Celestins. Lavoisier's house was in the Petit Arsenal, the entrance to which can be seen at the end of the Rue de la Cerisaie. The Bastille stands on the left with Porte-St.-Antoine adjoining. The Arsenal was bounded on the east by the *fosse* or moat, part of the defences of Paris. Le Mail, an avenue between the south wall of the Arsenal and the river, extends along an arm of the Seine, cutting off the Île St.-Louis and bridged by the Pont de Grammont. *Musée du Louvre, Plan de Turgot.*

While newly engaged on these heavy tasks, Lavoisier was struck down with grief by the sudden death of his father at his country house at Le Bourget in the autumn of 1775. Between father and son there had been the closest affection. "It is less the loss of a father that I mourn," he wrote, "than the loss of my best friend."

Upon his appointment to the Gunpowder Commission Lavoisier moved into residence at the Arsenal (Plate 18), where he remained until 1792. His laboratory there became for nearly twenty years a center of discovery and the rendezvous of leading men of science, including, among many others, Watt and Franklin.

XII

OXYGEN

AFTER LAVOISIER HAD ANNOUNCED TO THE ACADEMY THE results of his researches on mercury calx in April, 1775, he continued to consider this problem. His laboratory notebooks show that during 1776 and 1777 he repeated and studied Priestley's experiments. On May 10, 1777, as the Academy's volume of *Memoirs* for 1774 was still unpublished, he submitted to the Academy a revision of the memoir that he had read on November 12, 1774, on the calcination of lead and tin in sealed vessels; it was in this revised form that it finally appeared in the annual volume of *Memoirs* for 1774, which was published in 1778.

Air not a simple substance

THE GENERAL INTRODUCTION REMAINED MUCH THE SAME. Experimental details were now given for the calcination of tin. No figures could be given for lead, however, as the experiments had been difficult, because the calx of lead was volatile and settled on the walls of the vessel and therefore could not be taken out of the vessel and weighed. For tin, the results were clear. The total weight of the sealed vessel was the same before and after calcination. Gain in weight of the whole apparatus could be detected only after the sealed vessel was opened to admit air in place of that which had combined with the tin. This gain in weight was not the same as, but very nearly equal to, the gain in weight of the tin. It therefore appeared that the part of the air that had combined with the tin during its calcination had a density very nearly equal to that of the common air of the atmosphere.

So far, there was not much that was new except the experimental figures, but the statement about the density of the part of the air that had combined with the tin introduced a very significant change. Lavoisier now stated that other considerations, which, he said, would need too long and detailed an explanation at this stage, led him to think that the part of the air that had combined with the tin, and which combined with metals generally, was a little heavier than common air and that the part of the air left after calcination was a little lighter. The difference in their densities was, however, not very great.

Lavoisier had now arrived at the conclusion that it was only a part of the air, not the air itself in a purified state, that was capable of combining with the metals in calcination, while another part resisted this combination. The common air of the atmosphere was, therefore, not a simple substance, but a mixture of two very different constituents. Without anticipating the results of other experiments on mercury calx shortly to be described, Lavoisier asserted that the whole of the common air was not respirable; and that it was the respirable part that combined with the metals in calcination, the part remaining after the calcination being a kind of *mofette* or asphyxiant, incapable of maintaining the respiration of animals or the burning of combustible bodies. The common air about us was, therefore, composed of two different elastic fluids, one respirable and supporting combustion and combining with metals on calcination, and the other irrespirable and supporting neither combustion nor calcination.

Priestley's experiments had been given their true explanation by Lavoisier who had realized, after much experiment and much thinking, the true significance of Priestley's isolation of the substance mistakenly named by him "dephlogisticated air."

"Eminently respirable air"

FURTHER ADVANCE CAME IN 1778. ON AUGUST 8 OF THAT year, Lavoisier read to the Academy a revised version of his memoir of April 26, 1775, on mercury calx. This new account appeared in the annual volume of the *Memoirs* for 1775, which was published in 1778. He now stated that the principle that combined with the metals during their calcination and increased their weight was "nothing other than the most salubrious and purest part of the air." The numerical results were identical with those given in 1775, but the explanation differed.

The air obtained by heating mercury calx alone was "purer than the very air in which we live." Combustibles burned in it with a much greater brilliance than in common air; charcoal burned in it with a brilliance almost like that of phosphorus. It consumed combustibles with astonishing rapidity.

As in 1775, so in 1778, Lavoisier reported that when mercury calx was reduced with charcoal fixed air was given off. However, he now concluded that, since the charcoal was consumed in this change, the fixed air produced was due to the combination of the charcoal with "the pure part of the air." Black's fixed air was, therefore, a compound of charcoal with "pure air," which Lavoisier now began to call "eminently respirable air."

With this conclusion, the problem of the calcination of metals in air, their resultant gain in weight, and the production of fixed air when they were reduced with charcoal, was solved. Metals combined with eminently respirable air, one of the two constituents of common air, to form calces; and, in the reduction of their calces with charcoal, the eminently respirable air contained in the calces combined with the charcoal to form fixed air, or, in modern terms, carbon dioxide.

Priestley's experiments had been forged into a weapon which, although not yet fitted to overthrow the phlogiston theory, already threatened its supremacy in chemical philosophy. The new explanations given by Lavoisier failed, however, to attract any supporters for some years. The older theory was too widely accepted; and chemists had been so long accustomed to think within its framework that they found it too difficult as yet to reject their accepted beliefs.

At once Lavoisier proceeded to revise and extend the conclusions that he had made earlier on the results of other experiments. Phosphorus in burning, he was now able to assert, combined, not with air, but with the purest part of the air, to form "acid of phosphorus." The other part of the air he now customarily termed *mofette.* (We now call it nitrogen.) Similarly, sulphur combined with the purest part of the air to form vitriolic acid. In quantitative experiments, he found that phosphoric and vitriolic acids were composed of more than half their weight of eminently respirable air. Nitric acid was also shown to contain this air, which appeared to enter into the composition of all acids.

A new theory of combustion

MEANWHILE, ON MAY 3, 1777, LAVOISIER HAD READ TO THE academy a memoir, published in 1780, on the respiration of animals. He first prepared mercury calx by heating mercury in a confined volume of air for twelve days. The air decreased by about one-sixth of its volume, and forty-five grains of calx were collected. The residual air asphyxiated animals and extinguished flames. He then heated the mercury calx in a separate apparatus and recovered the air that had been combined in the calx. On adding this recovered air to the asphyxiating residue from the previous calcination, he obtained air with the properties of the original common air; it supported combustion and respiration. Thus, he had separated common air into its two constituents and then reconstituted it by reuniting them. He demonstrated also that when animals breathed in air their respiration led to the formation of fixed air, as had been shown earlier by Black. Lavoisier noted, however, that two processes were probably taking place in respiration, first, that the respirable part of the air was changed into fixed air, and, secondly, that it combined in some way with the blood to give the latter its red color. The mephitic or asphyxiating part of the common air, the *mofette,* passed into the lungs and came out again unchanged, but the respirable part was exhaled as fixed air.

In another memoir, Lavoisier dealt with an explanation current among those who accepted the phlogiston theory and intended to show why a lighted candle was extinguished in a confined volume of air even though much air still remained. At that time it was held that the burning candle, like other ignited combustibles, gave off phlogiston and that the phlogiston was absorbed by the air surrounding the candle, as, to follow our previous analogy, a sponge absorbs water. It was believed that there was a limit to the amount of phlogiston that the air could absorb, as there was a limit to the amount of water that a sponge could absorb; therefore, when the air became saturated to that limit, it could absorb no more, and hence, for this reason and not because of lack of air, the candle flame was extinguished. The residue was therefore known as phlogisticated air. Lavoisier accordingly placed a lighted candle in eminently respirable air and found that, even when as much as two-thirds of the air had been converted into fixed air, the residue still supported combustion. This was a complete contradiction of current theory. It showed that when the flame of a candle burning in a confined volume of air was extinguished, it went out because it had consumed the eminently respirable air, or, as Lavoisier was soon to rename it, the *oxygen* contained in the sample of common air in which it had been confined.

Oxygen

AT THIS TIME, TOO, LAVOISIER EXPRESSED THE OPINION THAT pure air in passing into the lungs underwent some change in its conversion into fixed air, comparable to that which occurred in the combustion of charcoal. Since heat was set free in the conversion of charcoal into fixed air, as when charcoal or coal is burned in a grate or a stove, heat similarly set free in the production of fixed air in the lungs might be, when distributed by the blood through the body, the cause of the maintenance of the body at a constant temperature above that of its surroundings.

At the Academy on November 23, 1779, Lavoisier, after pointing to his conclusion that eminently respirable air was contained in all the acids, announced that he would give it the name of oxygen from the Greek words, ὀξὺς, acid, and γείνομαι, I beget. This marks the first appearance of a term that has since passed out of the world of science into the currency of daily speech.

Throughout the period of these fundamental scientific advances, involving much experiment, Lavoisier was heavily occupied, not only with the business of the Farm, but also with the duties of the Gunpowder Commission. He carried out some tests on the ash used by the saltpeter makers of Paris and recommended its replacement by an ash containing much more alkali which was actually available at a lower price. Naturally occurring saltpeter was reported in the chalk at Roche-Guyon and Lavoisier, accompanied by Clouet, another Commissioner, journeyed there to take samples. These were analyzed and saltpeter was detected, together with the corresponding salt of nitric acid with lime as a base (calcium nitrate). Methods of extraction were prescribed and suitable details for the erection of artificial niter beds in the district were given. For three months in 1778 Lavoisier prospected widely in the provinces of France in search of natural deposits of saltpeter to increase the national supply, an objective which he tirelessly pursued. Experiments were made at the Arsenal under his direction.

The whole administration of the powder supply had been overhauled. Every detail was studied. Staff members were carefully chosen, and candidates for appointment were examined in chemistry and mathematics and the necessary technical knowledge. The powder was already improving in quality and its range was greater. In all this work Lavoisier took the lead. His organizing ability, his skill in conference and discussion, and his rapid drafting of reports and instructions, were invaluable.

As a relaxation from his many arduous duties, Lavoisier spent the vacations of the Academy in the country. At Le Bourget he had inherited a house from his mother's family. In 1778 he bought the château and estate of Fréchines, between Blois and Vendôme. Here he was able to develop his interest in agriculture; and here he began field experiments on a large scale for the general improvement of farming and of the lot of all those engaged in it.

XIII

HEAT AND RESPIRATION

LAVOISIER, AS WE HAVE SEEN, HAD BEEN INSPIRED BY BLACK'S researches on the alkalis. He had closely followed the method used by Black in tracing the detail of chemical changes by using the balance at every stage. His experiments on the alleged conversion of water into earth and the whole of his work on combustion reveal the source of the method that he employed so successfully.

Latent heat

BLACK HAD MADE ANOTHER DISCOVERY IN 1761 WHICH greatly interested Lavoisier. He had pointed to the common observation that, when ice was melting, it melted slowly and not all at once, and that snow accumulated during the winter thawed into water over a long period and needed even the sun of summer to cause its final disappearance from the mountains. Black had therefore concluded that ice and snow in melting took in large quantities of heat. If they did not, he pointed out, the effect of a thaw would be sudden, and great masses of ice and snow would change into water at once with catastrophic effects. But it was commonly believed that when ice or snow had reached its melting point only a further small addition of heat was necessary to convert it entirely into water. Black showed, however, that when a mixture of ice and water was placed in a warm room, the temperature of the mixture, as recorded by a thermometer, remained constant, at the melting point of ice, so long as any ice remained unmelted. Yet, during all this time, heat must be passing into the mixture and serving to melt the ice, and not to heat the water, since its temperature remained unchanged. When all the ice was melted into water, the temperature of the water then began to rise until it equalled that of the surroundings, that is, the temperature of the air in the room in which the mixture of ice and water had been placed.

In a very simple way Black had been able to measure the heat that passed into the ice in the process of melting. He took two glass phials of the same size and into each of them he put five ounces of water. He froze the water in one of them by immersing it in a freezing mixture of snow and salt, and he cooled the other nearly to freezing point; he then placed both vessels, one containing ice at 32° F. and the other an equal mass of water at 33° F., in a warm room.

The temperature of the cooled water rose in half an hour to 40°, and therefore a quantity of heat had entered in that time sufficient to raise its temperature by 7°. This gave the rate of entry of the heat, 7° in half an hour, as Black put it.

The ice in the other vessel, equal in weight to the water, melted slowly and took 10½ hours to melt into water and to reach the same temperature as the cooled water, namely 40°F. It had taken 21 half-hours, that is 21 times as long. Therefore, 21 times as much heat had entered into the ice as into the cooled water. The heat absorbed corresponded therefore to 21 x 7°, or 147°, and of this 8° had been used in raising the temperature of the water produced by the melting of the ice from 32° to 40°. Therefore, 139° had been absorbed. We do not measure heat in this way now, but Black had proved satisfactorily that ice in melting absorbed a large amount of heat, in fact, about 80 per cent of the heat that would be necessary to raise an equal weight of water from freezing point to boiling point.

In 1762 Black went on to show that a similar absorption of a very much larger amount of heat occurred in the conversion of boiling water into steam. His discovery of this heat—which, with the heat that entered melting ice, he called "latent heat," or hidden heat, because its entry was not detectable on a thermometer, since it caused no rise of temperature—was applied with remarkable consequences by James Watt to the improvement of the steam engine by the invention of the separate condenser.

Lavoisier and Laplace

BLACK'S RESULTS HAD REVEALED A METHOD OF MEASURING heat by the amount of ice melted. Lavoisier, in collaboration with his friend, the mathematician Laplace (1749-1827), designed an apparatus, called an ice calorimeter, to measure heat in this way. The many experiments which they carried out, from 1782 to 1784, included measurements of the different quantities of heat evolved in a number of chemical changes. These measurements are historic because they mark the foundation of what is now known as the science of thermochemistry, the practical application of which has since become of immense importance in industry.

Other measurements with the ice calorimeter proved equally significant in the science of physiology, since they showed that respiration was a kind of combustion. These experiments were difficult to carry out and involved several separate determinations, but we shall summarize the various steps that led to their notable conclusion about the nature of respiration. It was known that the temperature of the animal body, for example, the human body in health, remains constant, although it is losing heat to its surroundings all the time. The problem was to find how the vital functions of the animal were able to maintain its temperature above that of its surroundings and restore the constant loss of heat.

In Lavoisier's previous work on respiration, he had already shown that the oxygen of the air was converted into fixed air in the lungs. As we have seen, he had concluded that the maintenance of the animal body at a constant temperature above that of its surroundings was due to the release of heat, or "matter of fire," as he called it, in the lungs in the conversion of oxygen into fixed air, in the same way as heat was given out when oxygen was converted into fixed air by its combustive action on charcoal. It was a simple explanation, not so elaborate in its mechanism as we now know the detailed process to be; but Lavoisier had no quantitative results to support his conclusion, and these he was now seeking in collaboration with Laplace.

The experimenters (1) put a guinea pig in the ice calorimeter for 10 hours and measured the resulting amount of ice melted, which, of course, measured the heat given out by the animal during that time; the heat evolved melted 13 ounces of ice. Then, (2), they determined, in a separate apparatus, the amount of fixed air produced by the combustion of a weighed quantity of charcoal; and (3) they measured, in the ice calorimeter, the heat produced by the combustion of another weighed quantity of charcoal. From (2) and (3) they could calculate the heat produced in the formation of any quantity of fixed air.

Next, (4), they determined how much fixed air was produced by the respiration of a guinea pig in 10 hours, the interval during which its loss of heat had been measured in (1). Calculation from (2) and (3) gave the quantity of heat evolved in the formation of this amount (4) of fixed air; it corresponded to the melting of 10½ ounces of ice. In the first experiment (1), when the guinea pig was placed in the ice calorimeter for 10 hours, the heat lost by the animal during that time had melted 13 ounces of ice, but Lavoisier and Laplace pointed out that there must be many unavoidable errors in such difficult experiments, and that these figures of 10½ and 13 ounces showed as much agreement as could be expected in the circumstances.

The heat given out by the guinea pig in the ice calorimeter in the period of 10 hours corresponded, they considered, to the heat produced in its respiration during that interval of time, because, while the guinea pig was in the ice calorimeter, it was, of course, losing heat all the time, and yet its temperature did not fall to the temperature of the ice, but remained throughout very nearly the same as it was at the beginning. The guinea pig's vital functions had therefore restored the heat that it was constantly losing to its surroundings. Over the period of 10 hours, experiment had shown that 13 ounces of ice were melted. Therefore, in that time a corresponding amount of heat had been restored to the animal by its vital processes, and almost the same amount of ice (10½ ounces) was melted by the separate formation of as much fixed air, from charcoal and oxygen, as that exhaled by the guinea pig in the same time. The heat given out, in the lungs as they supposed, in the conversion of oxygen into fixed air in the process of respiration was therefore the principal cause of the conservation of animal heat, that is, the preservation of the temperature of the animal body at a constant degree above the temperature of its surroundings.

Respiration a kind of combustion

"RESPIRATION," CONCLUDED LAVOISIER AND LAPLACE, "IS therefore a combustion, admittedly very slow, but otherwise exactly similar to that of charcoal; it takes place in the interior of the lungs, without the evolution of light, since the matter of fire set free is immediately absorbed by the moisture of those organs; the heat developed in this combustion is communicated to the blood which passes through the lungs and thence diffuses through the whole animal system."

In further comments, in which charcoal was named "the base of fixed air," they went on to say: "Thus the air that we breathe serves for two purposes equally necessary for our preservation: it takes away from the blood the base of fixed air, the excess of which would be very harmful; and the heat that this combination sets free in the lungs makes up for the continual loss of heat that we experience with regard to the atmosphere and surrounding bodies."

Considerable light was thus thrown on the vital process of respiration. The conversion of oxygen into fixed air, which Lavoisier and Laplace located in the lungs, we now know takes place in the muscles. The function of oxygen in respiration had, however, been revealed in these experiments in a convincing quantitative way; and it had been demonstrated for the first time that respiration was a chemical, not a mechanical, process, not a mere ventilation or cooling of the lungs by fresh air.

For the moment this work was carried no further. Later Lavoisier took it up again and began to study transpiration as well. He took Armand Séguin (1765-1835), a much younger man, as his assistant. But the Revolution came and publication of the work was much delayed. The experiments were long and difficult and the problem was complicated, but eventually, after Lavoisier's death, some of the later work was published by Séguin. It was this problem, however, that attracted Lavoisier's attention during the last years of his life. Two sketches (Figs. 26 and 27) by Mme. Lavoisier of the experiments in progress in the laboratory have come down to us. The work was cut short by Lavoisier's untimely death at the hands of the Revolutionary Tribunal. But we know that he had found that the consumption of oxygen by a human subject was greater during digestion and at low temperature and during movement, exercise, and work.

In conclusion, we quote from Lavoisier and Laplace the principle that they had established in their joint memoir on heat:

> When an animal is in a constant state and at rest; when it can live for a considerable time without inconvenience in the medium surrounding it; in general, when the conditions in which it is placed do not in any way sensibly change its blood and humors, so that after several hours the animal system shows no appreciable variation; the conservation of its heat is due, at least in great part, to the heat produced by the combination of the pure air breathed by the animal with the base of fixed air supplied to it by the blood.

XIV

PROTEAN PHLOGISTON

SO FAR LAVOISIER HAD NOT ATTACKED THE PHLOGISTON theory with any detailed critical analysis. He had pointed out on several occasions that his own theory of combustion offered explanations that were more in accordance with the facts of chemistry. In 1783, however, in a memoir entitled "Reflections on Phlogiston," submitted to the Academy and included in the annual volume of *Memoirs* for that year (not published until 1786), he exposed in a masterly analysis the many weaknesses of the accepted chemical philosophy. This is one of the classics in the literature of science and reveals in the highest degree Lavoisier's great powers of reasoning and exposition. It is one of the steep ascents by which we climbed to our present knowledge of chemistry.

Lavoisier began by stating that, in the series of memoirs that he had communicated to the Academy, he had reviewed the principal facts of chemistry, especially those relating to combustion, the calcination of metals and, in general, all chemical changes in which there was an absorption or fixation of air; and that he had explained all these changes in terms of a simple principle, pure air or vital air, which he had since renamed oxygen. This principle once admitted, the chief difficulties that beset chemical philosophy vanished and the facts were explained with an astonishing simplicity.

Phlogiston unnecessary to chemistry

IF, THEREFORE, ALL IN CHEMISTRY COULD BE SATISFACTORILY explained without recourse to phlogiston, it seemed very probable that phlogiston did not exist and that it was a merely gratuitous supposition. Moreover, it was one of the principles of sound logic not to multiply entities unnecessarily. Lavoisier pointed out that he could rely on negative proofs of this kind and content himself with showing that the facts were better accounted for without phlogiston than with it. "But," he said, "it is time for me to speak out plainly in a more precise and formal manner about an opinion which I look upon as an error that is fatal to chemistry and seems to have greatly delayed its progress by the harmful method of thinking that it has introduced."

All prejudice, he insisted, must be put aside. Facts must be seen only as facts; and all previous reasoning about them should be dismissed from the mind. Chemists, for the moment, must consider themselves as carried back to the days before Stahl and must forget, if it were possible, that he had ever devised a theory.

In Stahl's day, the main facts of combustion were as yet unknown. He had recognized in the process of burning only what struck the senses, namely, the evolution of heat and light. From the fact that some bodies burned and inflamed, Stahl had concluded that there existed in them an inflammable principle, which he had called "fixed fire" or phlogiston. He had not, however, always been very clear or simple in what he said about phlogiston and many modifications had been introduced, both by him and his successors. It would suffice to consider the theory as generally understood at this time.

If, however, Stahl had confined himself to this simple observation, he would scarcely have deserved the glory of becoming one of the patriarchs of chemistry and of being recognized as having effected a kind of revolution in that science. Nothing was more natural than to conclude that combustible bodies burned because they contained an inflammable principle. But we owed to Stahl two important discoveries which were eternal truths, independent of every system and of every theory: first, that the metals were combustible bodies and that calcination was a true combustion; and secondly, and more importantly, that the property of burning and of being inflammable could be transmitted from one body to another.

The first of these discoveries, said Lavoisier, was generally admitted. It necessarily led Stahl to recognize an inflammable principle in the metals; and, indeed, if combustion were due to the release of an inflammable principle fixed in combustible bodies, it must follow from the fact that metals were combustible that they contained an inflammable principle.

Stahl's second discovery, added Lavoisier, was of great significance. If, for example, charcoal, a combustible substance, were treated with vitriolic acid, an incombustible substance, the vitriolic acid could be converted into sulphur, a combustible substance. Thus the vitriolic acid acquired the property of burning while the charcoal lost that property. It was the same with the metals. By calcination, they lost their property of being combustible, but, if heated in contact with charcoal or with bodies that had the property of burning, they regained, at the expense of charcoal or such bodies, the property of being combustible. Stahl concluded from these facts that phlogiston, the inflammable principle or element, could pass from one body to another.

Calcination

PHLOGISTON, ACCORDING TO STAHL, HAD WEIGHT, AND HE even attempted to determine its weight. But, said Lavoisier, his theory did not account for a fact observed long ago, verified by Boyle and now indisputable, namely, that all combustible bodies increased in weight when they were burned or calcined. This was observed in a striking manner in the metals, in sulphur, in phosphorus, and so on. However, according to Stahl, phlogiston escaped from the metals when they were calcined and from combustible substances when they were burned; and since phlogiston had weight, these bodies should therefore lose weight instead of gaining it.

Chemists had evaded this difficulty as much as possible. An attempt had been made to combine two explanations by ascribing the gain in weight to the addition of fire in a free state, while still supposing that phlogiston was released from metals in calcination. This would mean that fire was a very heavy substance, since iron gained more than a third of its weight by simple calcination in the open air; and the loss due to the release of phlogiston, which also had weight, had to be allowed for, since it was replaced by the matter of fire. Fire must therefore be, if this explanation were true, a very heavy substance. But this was contrary to all facts; its weight was so small that it could not be made sensible by any physical test. This was proved in Lavoisier's own experiments; for instance, the very considerable amount of fire and heat released in the combustion of phosphorus showed no appreciable weight even with the most sensitive balance.

Further, far from there being any fire absorbed in the calcination of metals, there was on the contrary a considerable amount set free as, for example, in the calcination of iron or zinc in oxygen, in which both metals burned brilliantly. Above all, if it were correct that particles of fire were absorbed by the metals in calcination, it followed that, when the calcination was arranged to take place in hermetically sealed vessels, there would be an increase in the weight of the sealed vessels. It was found, however, that, if they were weighed, without opening them, before and after the calcination, no difference in their weight could be detected by even the most sensitive balances.

Lavoisier then went on to point out that he had previously shown that when metallic calces were reduced with charcoal, an elastic fluid was given off, similar in every respect to the fixed air contained in chalk and the mild alkalis. He drew attention to the many experiments that he had made which showed that sulphur and phosphorus and all combustible bodies in general gained in weight on combustion by combination with a part of the air in the same way as metals gained in weight on calcination by a similar combination.

Macquer's theory

THESE FACTS, LAVOISIER SAID, WERE DISCONCERTING TO THOSE who accepted the phlogiston theory. He observed that Macquer had tried to reconcile fact and theory by supposing that phlogiston, still admittedly an inflammable principle possessing weight, was the matter, not of fire, but of light. Fire and heat Macquer regarded as immaterial, light as material. This was preserving the name without saving the thing. It was not a defence of the phlogiston theory; it was an outrage on it. And it was not the theory of Stahl or of any other chemist except Macquer; it was Macquer's theory, not the phlogiston theory, but it demanded attention.

Macquer, in supposing that phlogiston was the matter of light and that it was contained in the metals and all combustible bodies and released from them in calcination and combustion, had agreed that, concurrently with the release of phlogiston, the metals and all combustibles combined with the purest part of the air, as shown in Lavoisier's extensive experiments. He had, in effect, apparently reconciled the phlogiston theory with the new facts. Closer scrutiny, however, contended Lavoisier, showed important contradictions with those facts. No appreciable weight, for instance, had ever been detected in light. Again, if phlogiston were the pure matter of light, all metallic calces ought to be converted into metals when heated in the focus of the burning lens, which concentrated the rays of light upon them, just as they were reduced by means of charcoal. This, however, did not happen. Metals, except gold, silver, and mercury, were calcined in the focus of the burning lens and their calces vitrified, while the same calces in contact with charcoal at a suitable heat recovered their metallic state. Therefore, the matter present in charcoal was not the same as what Macquer had presumed to exist in the rays of light; and so phlogiston was not the matter of light.

Macquer had tried to dispose of this evidence by asserting that calces could not be reduced to metals when they were in contact with air, because they were recalcined by the air. Lavoisier presented a decisive experiment to refute this explanation. Calces were placed in receivers filled with atmospheric *mofette,* the residual inactive part of air from which oxygen had been removed, and then subjected to the action of light by the burning lens. They were not reduced to metals. Thus light did not act in the same way as charcoal, and therefore phlogiston could not be the matter of light.

Contradictory properties of phlogiston

CHEMISTS HAD NOT, HOWEVER, ACCEPTED MACQUER'S THEORY in general; and they had resorted to special explanations in particular cases, so that phlogiston had been preserved in name only, and contradictory properties had been thereby unwittingly combined in one vague and ill-defined term. For example, when pure charcoal was burned in oxygen, the charcoal disappeared completely and the oxygen was converted into fixed air. If this was arranged to take place in a closed vessel, carefully weighed before and after the combustion, no gain or loss was found in the total weight. Inside the vessel, however, the oxygen had increased in weight in its conversion into fixed air by an amount exactly equal to the loss in weight of the charcoal.

Those who accept the phlogiston theory, said Lavoisier, when they are asked to explain what has happened in this experiment, are forced to recognize (1) that some of the matter of heat and light, that is phlogiston, is set free, escapes through the walls of the vessel, and is dispersed; and, since the total weight of the apparatus remains unchanged, they have to admit that this matter of heat and light, this phlogiston that has escaped, has no appreciable weight; and (2) that during the combustion a weight of fixed air is formed equal to the sum of the weights of oxygen and charcoal used, whence they have to admit that, independently of all theory, charcoal contains a heavy matter that cannot escape through glass vessels and which therefore is not the matter of heat and light. Thus, they are forced, in explaining the combustion of charcoal, to give the same name of phlogiston to two very different substances, to weightless matter that escapes through glass vessels, and to ponderable matter that combines with oxygen to form fixed air. Here, therefore, said Lavoisier, are two distinct substances which they confound, a phlogiston that lacks weight and a phlogiston that has weight; one is the matter of light, and the other is not. It was, he contended, by borrowing these properties, as one might say, now from one and now from the other of these substances, that they managed to explain everything.

In the same way these chemists had unconsciously recognized two kinds of phlogiston in the reduction of metallic calces. Gold, silver, and mercury, calces were reduced to the metallic state merely by heating without any addition. Each gave, on the one hand, the metal and, on the other, the oxygen that had been combined with it, the combined weight of oxygen and metal being equal to the weight of the calx before reduction. In this case, Stahl's followers explained that the matter of light from the red-hot charcoal of the furnace passed through the pores of the glass vessels in which the calces were heated and combined with the metal. But, since the combined weight of the metal and the oxygen was equal to the original weight of the calx, it followed that, if there were combined phlogiston in the metal, phlogiston had no weight.

On the contrary, however, consideration of the reduction of calces by direct contact with charcoal led to a very different conclusion about phlogiston. In reduction by this method, it was found that metal and fixed air were produced and that the total weight of the fixed air obtained was increased by the weight of the charcoal used. Therefore, in this case, the phlogiston possessed weight. Once more, the name phlogiston had been given to two very different substances, to the weightless matter of light and fire and to the ponderable matter of charcoal.

Against these chemists, continued Lavoisier, the reduction of calces with charcoal provided a further embarrassing argument. The substance that combined with a metal to form its calx was, beyond all doubt, oxygen, but it was released in the form of fixed air when charcoal was used in the reduction. Would they say that the phlogiston of the charcoal had combined with the oxygen to form fixed air? Certainly, the weight of the oxygen and of the charcoal used was found again in the fixed air. But, if the whole weight of the charcoal had entered into the fixed air, it had therefore not combined with the metal, or at least that which had combined with the metal had no weight, while that which had combined with the oxygen had weight. Again, there were two phlogistons with contradictory properties, one that possessed weight and combined with oxygen to form fixed air and one that lacked weight and combined with the calx to form the metal.

Lavoisier then brought forward a result that contradicted one of the fundamental applications of the phlogiston theory. It had been generally accepted by the supporters of that theory, that, as we have already seen, when a combustible was burnt or a metal calcined in air, the phlogiston released from the combustible or the metal combined with, or was absorbed by, the air to produce a residue of phlogisticated air. (The residual air was regarded as a kind of sponge, if we may make a comparison, saturated with phlogiston.) But, he now pointed out, if this combustion or calcination were carried out in oxygen, the oxygen could be used up to the last bubble without the formation of any of the phlogisticated air. (Returning to our comparison, this proved that there was no sponge left!) If the experiment were interrupted before completion, the unconsumed oxygen was entirely unchanged.

Enough has been given to show the detail of Lavoisier's destructive criticism. In summarizing it, we have used the term "oxygen," for the sake of clarity, although in fact Lavoisier used in this place the term "vital air," which, however, he often used as synonymous with oxygen. He went on to show that the phlogiston theory often contradicted itself; while phlogiston was said to be the cause of color, causticity, and taste, there were many substances, either highly coloured or caustic or with a pronounced taste, from which, according to the theory, phlogiston was absent.

A new chemistry

"ALL THESE REFLECTIONS," ADDED LAVOISIER, "CONFIRM what I have advanced, what I set out to prove, and what I am going to repeat again. Chemists have made phlogiston a vague principle, which is not strictly defined and which consequently fits all the explanations demanded of it. Sometimes it has weight, sometimes it has not; sometimes it is free fire, sometimes it is fire combined with earth; sometimes it passes through the pores of vessels, sometimes these are impenetrable to it. It explains at once causticity and non-causticity, transparency and opacity, color and the absence of color. It is a veritable Proteus that changes its form every instant!"

Lavoisier's destructive analysis had wrecked the whole basis of the phlogiston theory. He called for a new system in which theory was built on fact.

"It is time," he said, "to lead chemistry back to a stricter way of thinking, to strip the facts, with which this science is daily enriched, of the additions of rationality and prejudice, to distinguish what is fact and observation from what is system and hypothesis, and, in short, to mark out, as it were, the limit that chemical knowledge has reached, so that those who come after us may set out from that point and confidently go forward to the advancement of the science."

Setting out his own theory, Lavoisier stated that the general facts of combustion might be summarized as follows:

1. There is true combustion, evolution of flame and light, only in so far as the combustible body is surrounded by and in contact with oxygen; combustion cannot take place in any other kind of air or in a vacuum, and burning bodies plunged into either of these are extinguished as if they had been plunged into water.
2. In every combustion there is an absorption of the air in which the combustion takes place; if this air is pure oxygen, it can be completely absorbed, if proper precautions are taken.
3. In every combustion there is an increase of weight in the body that is burned, and this increase is exactly equal to the weight of the air that has been absorbed.
4. In every combustion there is an evolution of heat and light.

Concluded Lavoisier:

> My only object in this memoir is to extend the theory of combustion that I announced in 1777; to show that Stahl's phlogiston is imaginary and its exist-

> ence in the metals, sulphur, phosphorus, and all combustible bodies, a baseless supposition, and that all the facts of combustion and calcination are explained in a much simpler and much easier way without phlogiston than with it. I do not expect that my ideas will be adopted at once; the human mind inclines to one way of thinking and those who have looked at Nature from a certain point of view during a part of their lives adopt new ideas only with difficulty; it is for time, therefore, to confirm or reject the opinions that I have advanced. Meanwhile I see with much satisfaction that young men, who are beginning to study the science without prejudice, and geometers and physicists, who bring fresh minds to bear on chemical facts, no longer believe in phlogiston in the sense that Stahl gave to it and consider the whole of this doctrine as a scaffolding that is more of a hindrance than a help for extending the fabric of chemical science.

This was a new Lavoisier, militant, inspired, challenging the obsolete in science as his enlightened contemporaries in the *Encyclopédie* had challenged the obsolete in society, calling upon the defenders of an outworn theory to surrender, and raining down upon their heads a destructive fire of incontrovertible chemical fact.

XV

THE COMPOSITION OF WATER

ALTHOUGH LAVOISIER HAD THOROUGHLY DEMOLISHED THE foundations of the phlogiston theory, one problem in combustion still baffled him and prevented him from gaining supporters for his theory. Inflammable air had been isolated and studied carefully by Cavendish in 1766, as we have already seen. It was recognized as an air distinct from ordinary air; in fact, it was the second air to be so recognized, fixed air being the first. It was known to burn and its inflammability had given it its name; but so far, however, the product it gave on combustion was unknown. Since, according to Lavoisier, combustion was combination with oxygen, his inability to explain the combustion of inflammable air was a serious hindrance to the acceptance of his theory.

Combustion of inflammable air

AS EARLY AS 1774, HE HAD STUDIED THE BURNING OF INFLAMMABLE air in a simple experiment, in which he had ignited a jet of this air from a flask in which it had been produced by the action of dilute vitriolic acid on iron filings. He had weighed the flask and the acid contained in it and he had added a weighed amount of iron filings. He found that there was a decrease in the total weight, but he did not succeed in making any further observations. In 1775, by means of a candle flame, he exploded a mixture of common air and inflammable air confined in a vessel over water, and he observed that the inflammable air burned only in proportion to the common air introduced.

In 1776 or 1777, the precise date being now unknown, Macquer had studied the combustion of a jet of inflammable air. He held a piece of porcelain in the flame and observed that droplets of a colorless liquid, which seemed to be water, formed on it. It was water, but the significance of this important observation was not realized at the time and Lavoisier did not hear of it. He was experimenting in 1777 on inflammable air with another object in mind. Bucquet had told him that he thought that inflammable air on combustion gave fixed air, but, when Lavoisier made the experiment over limewater, he found that the limewater did not turn milky and therefore fixed air was not produced. Later, in 1781-82, he made some attempt with Gengembre to detect the acidity that should be produced when a jet of oxygen, the "acid-former," was burned in inflammable air and, of course, found none. Clearly, Lavoisier was much interested in this problem. While he could not have realized in those years the ultimate importance its solution might hold for his theory, he can scarcely be regarded as unconcerned about bringing this further instance of combustion within the range of that theory. He must often have asked himself, "What does inflammable air produce when it burns, that is, according to my theory, when it combines with oxygen?"

The solution of the problem came about in an indirect way. In 1781 Priestley had been experimenting on the explosion of inflammable air with common air in closed glass vessels by means of the electric spark. He noticed that after the explosion the inside of the glass vessel "became dewy." His friend, John Warltire (*c.*1738-1810), an itinerant scientific lecturer, had been discussing with him the possibility of finding out whether or not heat had weight by exploding a mixture of these two airs by the electric spark in closed glass vessels weighed before and after the explosion. But Warltire feared that the experiment might be dangerous, and that it would shatter the vessels. Priestley, that intrepid experimenter, however, had tried the explosion himself in glass vessels, while Warltire took Priestley's advice and tried it in copper vessels. Warltire observed that a "smoke" was expelled from the copper vessels after the explosion, when air was blown through to clear them for the next experiment; and he confirmed the observation of Priestley that glass vessels "became dewy."

Warltire's experiments, however, were made to find out whether heat had weight. Both Priestley and Warltire were interested in their discovery of a variable loss of weight, not in the formation of "dew" or "smoke." Accordingly, Priestley informed Cavendish of the curious loss of weight and Cavendish decided to repeat the experiments himself.

Henry Cavendish

HENRY CAVENDISH (1731-1810), WHOSE APPARATUS FOR THE collection of airs we have already mentioned, was an eccentric. His biographer, Wilson, described his character as "a blank" and "a series of negations."

> He did not love; he did not hate; he did not hope; he did not fear; he did not worship as others do. He separated himself from his fellow men, and apparently from God. There was nothing earnest, enthusiastic, heroic, or chivalrous in his nature, and as little was there anything mean, grovelling, or ignoble. He was almost passionless. All that needed for its apprehension more than the pure intellect, or required the exercise of fancy, imagination, affection, or faith, was distasteful to Cavendish. An intellectual head thinking, a pair of wonderfully acute eyes observing, and a pair of very skilful hands experimenting or recording, are all that I realize in reading his memorials. His brain seems to have been but a calculating engine; his eyes inlets of vision, not fountains of tears; his hands instruments of manipulation which never trembled with emotion, or were clasped together in adoration, thanksgiving, or despair; his heart only an anatomical organ, necessary for the circulation of the blood. [He was] a wonderful piece of intellectual clockwork.

Cavendish was the son of Lord Charles Cavendish, the third son of William, fifth Earl and second Duke of Devonshire. His mother was Lady Anne Grey, daughter of Henry, Duke of Kent. Among his many distinguished ancestors, he could count the redoubtable Elizabeth Hardwicke, "Bess of Hardwicke," Countess of Shrewsbury. In his younger days, his father seems to have kept him short of money and he learned such habits of economy that, when he became rich through inheriting several fortunes, he was unable to spend more than a fraction of his very large income. He loved to be alone and his shyness amounted almost to a disease; he devoted his life to scientific research. To his immense fortune he seems to have been completely indifferent; and he was almost as indifferent about the publication of some of his scientific discoveries, which remained unknown until they were found among his papers long after his death.

Priestley and Cavendish were both Fellows of the Royal Society. Their common interest in "airs" and in chemistry led Priestley to inform Cavendish of the result of the experiment that Warltire had been making on the explosion of inflammable air and common air, with the object of finding out whether heat had weight, and of the curious observation of a decrease in weight and of the formation of "dew." Cavendish was much interested in heat and refined measurement was one of the distinguishing characteristics of all his scientific work.

Cavendish's experiments

IN 1781 SHORTLY AFTER RECEIVING FROM PRIESTLEY INFORMATION about Warltire's results, Cavendish set out to repeat these experiments. He found that commonly there was no loss of weight, although this occasionally occurred, but even in these cases it was very small. He confirmed the observation about the formation of "dew" and decided to investigate the nature of this product. By exploding with the electric spark mixtures of inflammable air and common air in glass globes, he found that he could reduce the volume of common air by one-fifth, and that this was the maximum reduction that could be thus effected in its volume. His experiments showed that 423 measures of inflammable air were required to effect this reduction in 1,000 measures of common air. From this he concluded that the inflammable air and one-fifth of the common air had been "condensed into the dew which lines the glass." He made no suggestion that the "airs" had combined to form the "dew."

Cavendish then decided to collect the "dew" by experimenting on a larger scale. He left no diagram of the apparatus used in this historic experiment, but his description is as follows:

> The better to examine the nature of this dew, 500000 grain measures of inflammable air were burnt with about 2½ times that quantity of common air, and the burnt air made to pass through a cylinder eight feet long and three-quarters of an inch in diameter, in order to deposit the dew. The two airs were conveyed slowly into this cylinder by separate copper pipes, passing through a brass plate which stopped up the end of the cylinder; and as neither inflammable nor common air can burn by themselves, there was no danger of the flame spreading into the magazines from which they were conveyed. Each of these magazines consisted of a large tin vessel, inverted into another vessel just big enough to receive it. The inner vessel communicated with the copper pipe, and the air was forced out of it by pouring water into the outer vessel; and in order that the quantity of common air expelled should be 2½ times that of the inflammable, the water was let into the outer vessels by two holes in the bottom of the same tin pan, the hole which conveyed the water into that vessel in which the common air was confined being 2½ times as big as the other.
>
> In trying the experiment, the magazines being first

> filled with their respective airs, the glass cylinder was taken off, and water let, by the two holes, into the outer vessels, till the airs began to issue from the ends of the copper pipes; they were then set on fire by a candle, and the cylinder put on again in its place. By this means upwards of 135 grains of water were condensed in the cylinder, which had no taste or smell, and which left no sensible sediment when evaporated to dryness; neither did it yield any pungent smell during the evaporation; in short, it seemed pure water . . .
>
> By the experiments with the globe it appeared, that when inflammable and common air are exploded in a proper proportion, almost all the inflammable air, and near one-fifth of the common air, lose their elasticity, and are condensed into dew. And by this experiment it appears, that this dew is plain water, and consequently that almost all the inflammable air, and about one-fifth of the common air, are turned into pure water.

The mysterious "dew" had therefore proved to be "plain water." The inflammable air and one-fifth of the common air had "condensed" to form this water and Cavendish made no suggestion of chemical combination between the two airs.

These results were not published at the time of their completion. Cavendish had found that, when he used Priestley's dephlogisticated air (Lavoisier's oxygen) instead of common air, the resultant water contained nitric acid. He turned aside to investigate this and incidentally discovered the composition of nitric acid. He further noted that, on passing the electric spark through common air, dephlogisticated air being added in sufficient quantity to combine with all the phlogisticated air, a small intractable bubble of air was left that could not be further reduced. A century later Rayleigh and Ramsay showed that this residue was an inert gas, to which they gave the name "argon."

Cavendish had turned aside to study the formation of nitric acid, but when his memoir was read to the Royal Society on January 15, 1784, the results of his work were already known to most of his contemporaries. However, for the moment, we turn our attention to the explanation that Cavendish gave of the formation of water.

He regarded inflammable air either as phlogiston itself or water united to phlogiston, inclining to the latter alternative; and dephlogisticated air was water deprived of its phlogiston. Set out in a simpler way, Cavendish's explanation appears as follows:

$$\begin{array}{ll} \text{Dephlogisticated air} & = \text{Water} - \text{Phlogiston} \\ \text{Inflammable air} & = \text{Water} + \text{Phlogiston} \\ \hline \end{array}$$

And, by "condensation,"

$$\underline{\text{Dephlogisticated air} + \text{Inflammable air} = \text{Water}}$$

On exploding the mixture, condensation occurred, in Cavendish's view, and water was obtained from each of the airs. Evidently, in the reaction between the two airs, the excess of phlogiston in the one was supposed to make up for the lack of it in the other; and water was produced from both, each of the airs contributing its constituent water, and the water was not the result of the chemical union of the two airs. Cavendish had no idea that water was a compound of dephlogisticated air and inflammable air. Water, he considered, existed as water in each of the two airs that had been exploded or ignited in the mixture and was condensed or deposited from each of them.

Lavoisier's experiment

CAVENDISH'S MEMOIR WAS, AS WE HAVE ALREADY NOTED, read before the Royal Society in January, 1784, but the experiments had been made in 1781. In June, 1783, Charles Blagden, Cavendish's assistant, visited Paris, where he called upon Lavoisier and in conversation gave him an account of Cavendish's experiments, as was usual among men of science. It was said later by Blagden that at first Lavoisier had some difficulty in believing that inflammable air on combustion yielded water. We know that he had expected the product to be acidic. Lavoisier's doubts are not surprising, because he must have realized the immense significance for his theory of the unexpected nature of the information that had been imparted to him, if it were correct. It answered the question that had long bothered him, "What does inflammable air produce when it burns, that is, according to my theory, when it combines with oxygen?"

On June 24, without delay and while Blagden was still in Paris—and, indeed, present at the experiments—Lavoisier made a hurried and improvised attempt in collaboration with Laplace to verify Cavendish's results. On the very next day, June 25, 1783, he made a brief report of his results, which he communicated to the Academy, naming Laplace as his collaborator and stating that the combustion of inflammable air with oxygen carried out in a closed vessel yielded "water in a very pure state."

Following this, on November 12, 1783, at a public meeting of the Academy, Lavoisier read a memoir on these experiments entitled "On the Nature of Water and on Experiments that appear to prove that this Substance is not properly speaking an Element, but can be decomposed and recombined." The memoir was printed almost at once in the December issue of Rozier's journal. Later it appeared in a more extended form in the annual volume of the *Memoirs* of the Academy for the year 1781, not 1783 as might have been expected, since the volume for 1781, with characteristic delay, did not appear until 1784.

In the account given in Rozier's journal, only slight reference was made to Cavendish's as yet unpublished work and it was made to appear, in the few lines referring to it, as a mere confirmation of the experiment of Lavoisier and Bucquet with the added observation of the formation of "dew." It was also described as a further confirmation of some experiments by Monge, of which the Academy had been informed after Lavoisier had made his brief report in the previous June. Monge had burned known volumes of the two airs and had obtained an amount of pure water very nearly equal in weight to the combined weight of the two airs used.

The experimental detail described very briefly how Lavoisier and Laplace had burned measured volumes of the two airs in jets in a glass receiver over mercury. Droplets of water formed on the glass and drained on to the surface of the mercury. This water was collected and weighed. It weighed nearly as much as the sum of the weights of the two airs used. The method of experiment and the quantitative results were, alike, far inferior to those of Cavendish.

Water composed of inflammable air and oxygen

BUT THE CONCLUSION MADE BY LAVOISIER WAS OF FAR greater importance. Water, he said, must be composed of inflammable air and oxygen. This is the first assertion in chemistry that water is a compound of the two gases that we now know as hydrogen and oxygen. Lavoisier, and Lavoisier alone, had made the correct deduction from Cavendish's experiments and he is entitled to every possible recognition for it. Cavendish had shown that the combustion of inflammable air yielded water; Lavoisier concluded that water was a compound of inflammable air and oxygen.

Lavoisier added that his conclusion was so different from current opinion that he proposed to multiply proofs and, especially after establishing the nature of water by compounding it from its constituent parts, to apply himself to decompose it into those constituents.

For this latter purpose, he made a further experiment. He filled a glass bowl with mercury and inverted in it another vessel filled with mercury; into this second vessel he introduced a small amount of water and some unrusted iron filings. The metal gradually rusted by combination with the oxygen of the water. At the same time a quantity of inflammable air was set free from the water proportional to the amount of oxygen that combined with the iron; the weight of oxygen was determined by drying the residual and now rusted iron and noting its gain in weight, and the weight of inflammable air was obtained by measuring its volume and thence calculating its weight. Here, said Lavoisier, water had been decomposed into oxygen, which had combined with the iron and converted it into the rust or calx of iron, and inflammable air, which had been set free. These two substances, by the previous experiments, could be combined to form water; and water was, therefore, not a simple substance, not an element, as it had so long been held to be.

This experiment, together with Cavendish's experiment, which he had hurriedly and very imperfectly repeated, led Lavoisier to the conclusion that we still accept today. Earlier in his chemical researches, he had said that if he were reproached with having borrowed proofs from the work of others, "It will at least not be disputed that the conclusions are my own." While there must always remain among the admirers of Lavoisier's genius a feeling that he did not give full recognition to Cavendish's work on this problem, justice cannot refuse to recognize that it was Lavoisier who drew the correct conclusion from that work and thereby discovered the composition of water. He alone of all the chemists of his time saw through the confusions of the phlogiston theory and realized the significance of the results obtained by his English contemporary.

We are still, however, in the year 1783, and Cavendish's work had not been published. On January 18, 1784, Cavendish read his memoir to the Royal Society and it appeared in the volume of the *Philosophical Transactions* for that year. Blagden had in the meantime become Secretary of the Royal Society and accordingly inserted in the text of Cavendish's memoir before it was printed a statement that he had informed Lavoisier of Cavendish's experiments in 1783. There was possibly some feeling on the matter that Lavoisier might have used the information given to him by Blagden to anticipate the publication of a discovery made by Cavendish.

The history of the discovery of the composition of water bristles with difficulties and with personalities. It has been the cause of much bitter controversy, but the facts are plain and the controversy now seems to have been scarcely necessary. James Watt also made through his friends a claim to have discovered the composition of water, but his ideas were confused and rendered the issue even more complicated. Watt has no claim here. Cavendish obtained the first experimental result and regarded the water as previously present in both airs. Lavoisier said that the water was formed by the combination of the two airs and confirmed his explanation by decomposing water into its two gaseous constituents.

In the extended version that Lavoisier gave in the memoirs for 1781 (published in 1784), the same indifferent notice was taken of Cavendish's work and Lavoisier's repetition of it was no more satisfactory than in the earlier version in Rozier's journal. However, some new and interesting experiments were described by Lavoisier and Meusnier, in which a weighed amount of water was admitted, drop by drop, at the upper end of a sloping iron gun barrel heated red-hot at its middle. The undecomposed fraction of the water was condensed at the lower end and weighed, while the gaseous product, inflammable air, of the fraction decomposed was collected over water and measured. The outer surface of the iron was calcined in the heating, however, which prevented an exact determination of the increase in weight of the iron by its combination, on its inner surface, with the oxygen set free from the water; this was not possible even when the outer surface was protected by coatings of clay and lute.

In other experiments separately made to find out what metals would decompose water in this way, Lavoisier found that copper at a red heat did not react with water. He then used a copper tube filled with small pieces of iron in foils and spirals. The copper was unattacked; the water was decomposed, its oxygen combining with the iron and its inflammable air being collected as before. Calculation from the experimental results showed that one part by weight of inflammable air had combined with about 6½ parts by weight of oxygen. The result is not a good one, the values being as 1 to 8, but the important end in view had been attained, namely, the proof that water was decomposed into inflammable air and oxygen. Analysis had completely confirmed synthesis.

Acceptance of Lavoisier's theory

ELUCIDATION OF THE COMPOSITION OF WATER SOLVED A problem which had long been a stumbling block to Lavoisier's theory and had prevented his contemporaries from sharing his views. When a metal was dissolved in the solution of an acid in water, inflammable air was evolved and a salt was obtained on removal of the water by evaporation of the residual solution to dryness. But when the calx of that metal was used in this experiment, in place of the metal itself, only the salt was obtained and there was no evolution of inflammable air. Those who accepted the phlogiston theory, and this means chemists in general, had a simple explanation for these differing results: inflammable air was phlogiston; in the first of these reactions, it came from the metal, whereas, in the second reaction, where the calx was used, there was no phlogiston because it had been lost by the metal on conversion of the latter into calx, and therefore there was no evolution of inflammable air. The salt, they considered, was produced by the combination of the calx with the acid. So much was this the accepted opinion that when Cavendish had originally prepared inflammable air in 1766 by the solution of metals in dilute acids and named it, he said of these metals that "their phlogiston flies off . . . and forms the inflammable air."

The solution of a metal in an acid with the formation of a salt and the evolution of inflammable air, which we may represent:

metal + acid + water = salt + inflammable air,

may be set out according to this explanation:

$$\underbrace{(\text{calx} + \text{phlogiston})}_{\text{metal}} + \text{acid} = \underbrace{(\text{calx} + \text{acid})}_{\text{salt}} + \underset{\text{(inflammable air)}}{\text{phlogiston}}$$

The part that the water might play in these changes was ignored and its function was unknown; it was looked upon as nothing more than the medium in which they occurred.

So far Lavoisier's theory had not been able to provide any better explanation than that given in terms of the phlogiston theory. Indeed, it could provide none at all. Now, however, there was at hand an explanation in terms of his theory which took account of the part played by the water. The inflammable air had come from the water, the oxygen of which had converted the metal into a calx, which had then combined with the acid to form a salt. Thus, it was the decomposition of water that produced the inflammable air. The change:

$$\text{metal} + \text{acid} + \text{water} = \text{salt} + \text{inflammable air},$$

could now be set out:

$$\begin{aligned} &\text{metal} + \text{acid} + \text{water} \\ &= \text{metal} + \text{acid} + \underbrace{(\text{inflammable air} + \text{oxygen})}_{\text{water}} \\ &= (\text{metal} + \text{oxygen}) + \text{acid} + \text{inflammable air} \\ &= \text{calx} + \text{acid} + \text{inflammable air (since metal + oxygen = calx)} \\ &= (\text{calx} + \text{acid}) + \text{inflammable air} \\ &= \text{salt} + \text{inflammable air (since calx + acid = salt).} \end{aligned}$$

The problem now stood explained and the long-standing objection to Lavoisier's theory was removed. When zinc, for example, dissolved in vitriolic acid diluted with water, the zinc combined with the oxygen of the water to form calx of zinc, which then combined with the acid to form a salt, while the inflammable air of the water was liberated and evolved. Soon afterwards chemists began to accept the new theory; the main objection to it had been removed by Lavoisier's discovery of the composition of water.

(Acids were then regarded, on Lavoisier's theory, as compounds of certain substances, such as phosphorus and sulphur, with oxygen. Today we regard the acids as the further combinations of these compounds with water, and the inflammable air, or hydrogen, as evolved from the water that has combined to form the acid, which is the same thing as saying that it comes from the acid.)

XVI

THE SERVANT OF THE STATE

The Tax-Farm

LAVOISIER HAD, AS WE HAVE SEEN, ENTERED THE TAX-FARM in 1768 as assistant to the Farmer-General Baudon, having bought a one-third share of Baudon's interest. As he compiled an account of the Farm's returns over many years, from some of the earliest leases up to 1774, and therefore for leases both before and after he became one of its officials, we have some very valuable and informative figures about his own profits and those of others.

The annual income of a Farmer-General of Taxes naturally varied from lease to lease, according to the varying terms of the lease and the fluctuations in national trade and prosperity. Under the lease of Henriet (1756-62), it amounted to 40,000 livres on an advance of 1,000,000 livres, under that of Prévost (1762-68) to 55,000 livres on an advance of 1,200,000 livres, and under that of Alaterre (1768-74) to slightly more. All these were much below the sums alleged by rumor during the Revolution; the profit was generally about 4 to 6 per cent. When Lavoisier first entered the Farm in 1768, he had to advance

520,000 livres for his one-third share in a place, and he would receive accordingly one-third of the annual income of a Farmer-General, which, as will be clear from the figures given above, was not an excessive profit. In 1771, he acquired a half-share, as Farmer-General Baudon was growing old and wished to have less work on his hands, and therefore Lavoisier had to advance a further 260,000 livres. Under the lease then current, that of Alaterre (1768-74), Lavoisier's profits probably amounted to 100,000 livres. Under the new lease, that of Laurent David (1774-80), Lavoisier again had a half-share in a place.

Twenty days after the signature of the lease for a sum of 162 million livres to be provided annually for the Government, the Minister, the Abbé Terray, informed the Farmers-General that he had reserved a share of the profits for the King, a condition that was not included in the terms of the lease. The Farmers, after announcing that they would withdraw from their contract, were forced to yield by a threat from Terray that the funds they had advanced would in that case not be returned to them except as life annuities. Though they were apprehensive about their profits under the lease, their fears were not justified by the event, because Turgot soon replaced Terray and improved the administration, and because the vintages for this period were exceptionally heavy, so that the lease brought each of the Farmers a yearly income of more than 100,000 livres on an advance of 2,700,000 livres.

Although Turgot was eager to introduce reforms, he was not able to do all that he wished. He drew Louis XVI's attention to the abuses by which pensions and incomes for favorites and other parasites had to be provided from the resources of the Farm, in fact from the profits of the Farmers-General; but he could not obtain authority to cancel those instituted by the late King Louis XV. Turgot was, however, able to make two badly needed changes. First, he provided that places as Farmer-General and Assistant Farmer-General, often given in the past by favor to relatives of these officials or to other incompetents, should in future be reserved for those who had proved themselves capable, by service in the Tax-Farm, of discharging their duties effectively and of being useful to the administration. Second, he banned the introduction of any new pensioners and parasites, "riders" on the Farm, as they were popularly called, who had in the past been allotted a place or a share of a place with the corresponding income but without any duties; as such places as were then held by these "riders" became vacant, they were to be distributed among those Farmers-General and Assistant Farmers-General who had only a share in and not the whole of a place.

Under this new lease, Lavoisier took a more active part in the work of the Farm. Under the previous one, he had been a member of two committees, but now he was appointed to several others—the tobacco committee for Paris, which met twice every month, the committee for import duties for Paris, which met every Wednesday, the salt-works committee for Franche-Comté, Lorraine, and Les Trois Évêchés, the committee on administration, and the committee on personnel. At the same time he was also appointed correspondent for the duties on exports, salt, tobacco, and taxes for the city of Paris and its suburbs; and he had many journeys to make, going, for instance, to Brittany in 1778 on the Farm's affairs.

Baudon, Lavoisier's associate, died in 1779 and Lavoisier became full Farmer-General in October of that year. On receiving the appointment, he had to pay the usual fee of 25,000 livres; promotions were subject to payment, as they were made under the seal of the chancellery. A condition of his appointment was that he should set aside a third of the profits for the remaining years of the lease for Baudon's widow, with whom he made an agreement for a fixed sum calculated on the profits of the years of the lease already run. When the actual profits for 1780 were known, in 1787, and were found to be much greater than expected and estimated when the agreement was made, he generously sent to Mme. Baudon a third of the extra amount realized.

When David's lease expired in 1780, a new one was drawn up in the name of Nicolas Salzard. But Necker, the Genevan banker (who would be dismissed and recalled, and again dismissed and recalled in the difficult and troubled times that lay ahead, until his final resignation and flight in 1790), was now Minister, and the terms of the new lease marked great changes to the advantage of the State. The number of Farmers-General was reduced from sixty to forty; the office of Assistant Farmer-General was suppressed; and the profits of the Farm were to be shared equally between the Farm and the Treasury, as soon as the sums collected exceeded by 3,000,000 livres the price fixed in the lease. Interest on the advance, now restored to 1,200,000 livres for each Farmer-General, was fixed at a flat rate of 7 per cent; while allowances were raised, to include office expenses and gratuities, to 32,000 livres. From the moment when the Farmers-General of Taxes were called upon to divide their profits with the State, their organization became an administration, admittedly self-interested, but an administration rather than a finance company, and began to tend towards modern fiscal systems. Despite these changes, the share of the Farmers-General was still high, thanks to the suppression of "riders" and pensioners, and to the reduction of the number of Farmers, for each of whom it now stood at the annual figure of 75,000 livres.

Lavoisier was chosen to be one of the forty Farmers-General in the new system. His abilities, evident by his previous work in the Farm and as Commissioner for Gunpowder, were too well known and too highly valued by the Ministers concerned and by his colleagues not to be used in their service. In 1783, d'Ormesson, who had then become Minister, appointed him to the committee of administration, the most important of the Farm's committees, as it transacted business with the Government. D'Ormesson's letter to Lavoisier on this appointment, said: "You have long known my views. I take this opportunity of affording you proof of them by informing the Farm of my wish that you should be appointed to the committee of administration. I believe that they will readily acquiesce in my wishes, as I am well aware of the respect and consideration in which you are held by them." Lavoisier was also especially entrusted with the higher direction of the import duties for Paris and the accounts of the salt works of the three districts; he had been appointed to these committees during the term of the previous lease.

As early as 1774, he had pointed out that by establishing a uniform customs duty throughout France, expenses would be reduced by 4%, through the abolition of the brigades of internal customs officials and of more than a thousand customs offices, "independently," he added, "of the increase in production that would be the natural consequence of uniform taxation, even if the provinces were granted, through the resulting decrease of the *taille* [the tax on personal property] and other duties, a more considerable indemnity than they would have any right to expect."

Smuggling would likewise become less attractive to those tempted to profit by this crime. One of the chief difficulties of the Tax-Farm was the prevention of smuggling, not merely across the frontiers of France, but over the boundaries between her provinces, since the duty on some commodities varied considerably from province to province. Salt, for instance, cost almost thirty times as much at Angers as it did at Nantes; the amount of salt that every subject had to purchase annually from Government supplies ranged from nine to fourteen livres in different provinces, while in some parts, relieved under an old law, the amount was quite small. The system encouraged smuggling, and there was a whole army of smugglers—men, women, and children, and soldiers too—because contraband was highly profitable. In one year alone, there were more than a thousand arrests and 2,700 seizures of contraband in houses. The severe law of Louis XIV inflicted the penalty of death on cavalrymen convicted of smuggling salt, and, even when it was mitigated, three hundred smugglers were condemned every year to the galleys, where they formed a third of the convicts.

Officials of the Farm, having the right of domiciliary entry, were often led to commit abuses which the central administration had difficulty in repressing because this might make the officials less concerned with discharging their duties and thus cause a reduction in the yield of revenue. All this increased the odium in which the Farm was held. There was a right of appeal to a court of taxes, but it was merely illusory. When Necker was Minister, Lavoisier was charged with the work of drafting and editing the instructions given to the officials of the Farm, to encourage those who were not conscientious in the discharge of their duties, and to repress others who multiplied prosecutions and made themselves and the Farm hated by their exactions.

During part of his time in the Farm, administration of the district of Clermontois was in Lavoisier's jurisdiction. Here he relieved the Jews of an odious toll, called *pied-fourchu* or cloven hoof, the inhuman suggestion of which is only too obvious, and which, to the shame of the Governments concerned, survived elsewhere much later. Lavoisier succeeded in securing its abolition. As an expression of gratitude, the Jewish community of Metz sent a deputation to Lavoisier to thank him and to offer him some Passover cakes as a token of religious fraternity.

Inside the Tax-Farm, Lavoisier made many improvements. He undertook a study of the consumption of wood in the salt works, for which he received the congratulations of d'Ormesson, and he analysed the salt from various sources. His administrative and reforming work seems to have met with approval everywhere; he showed himself dissatisfied with the wastefulness of the Farm's system, but it was the official system of his country, and he did his best to improve it, always with the intention of achieving equity and of making the burden of taxation lighter for the poorer classes.

One of his reforms, however, met with violent criticism through ignorant misrepresentation. It occurred during the period in which he was especially concerned with the import duties for Paris. At this time it was estimated that one-fifth of the merchandise entering the city was brought in as contraband by enterprising smugglers, to the detriment of the Tax-Farm, and therefore of France, and also of the honest traders, who complained of the connivance in this crime of their less scrupulous competitors. To remedy this, Lavoisier proposed to the Government the construction of a wall around the city, but no action was taken for two years until Calonne, who was Minister from 1783 to 1787, resolved to carry out the proposal. Calonne unfortunately authorized the architect Ledoux to build the wall; he erected sumptuous buildings at the customs barriers and spent thirty million livres. Naturally, there was much criticism and much misrepresentation. The customs offices were said to be new fortresses for the subjection of the citizens; the wall was said to be designed by the Farmers-General to cut off the fresh air from Paris and to prevent the escape of the deadly exhalations constantly discharged into the atmosphere of the city. Pamphleteers were busy, and they put the blame, not on the Farm, but on Lavoisier. The Duke of Nivernais, Marshal of France, bluntly expressed the soldier's opinion that the originator of the scheme ought to be hanged, and wits coined the phrase *le mur murant Paris rend Paris murmurant*. While Ledoux merited every criticism for his foolish extravagance, a measure of strict justice was made to appear as the arbitrary action of Lavoisier, and the ignorant pamphleteers misrepresented as the worst enemy of the people one who was devoted to the interests of his fellow citizens.

The lease of Nicolas Salzard terminated in 1786; under it the annual profits of a Farmer-General had amounted to 75,000 livres. It is estimated that from 1768 to 1786 Lavoisier's profits in the Farm amounted to 1,200,000 livres. A new lease was drawn up in 1786 in the name of Jean Baptiste Mager, but it did not run for the full term, because after the outbreak of the Revolution the Tax Farm was suppressed by the National Assembly in 1791. The liquidation was entrusted to six of the Farmers-General; Lavoisier was not among these and so his connection with the Farm ceased, although it was yet to be the cause of his lamentable end.

The Gunpowder Commission

WHEN THE PHYSIOCRAT TURGOT, DURING HIS BRIEF PERIOD OF office from 1775 to 1776 as Controller-General of Finance, initiated the reform of the system by which the nation was supplied with gunpowder, it was by Lavoisier's hand, as we have already seen (p. 125) that the reform was effected. Quality and output were increased and costs were reduced. The annual national production of saltpeter rose in seven years from one and a half million to two and a half million pounds, and to more than three and three-quarter million pounds in 1788. Fifteen hundred men were employed in the manufacture of a product previously bought from foreign sources, and retained, in the form of wages, in the provinces of France money that would otherwise have gone abroad. In this year of 1788, the amount of gunpowder stored in the magazines was five million pounds, while the saving in money to the nation was estimated at twenty-eight million livres. Before 1775 the range of French military gunpowder had been from 136 to 155 meters; five years later it was from 224 to 253 meters and its superiority, Lavoisier noted, had become a matter of concern to the English. Production was increased to such an extent that France was able to export gunpowder, supplying, among others, the American colonists in their War of Independence. It is no farfetched conclusion that the new kind of mobile war that was soon to be waged by Napoleon owed much to the labors of Lavoisier as organizer of the Gunpowder Commission. In the prosecution of his researches on this problem, an economic as well as a scientific one, he travelled widely in

France, spending three months in Touraine and Saintonge, to test and analyze reported natural sources of saltpeter. From 1775 to 1783 he carried out numerous experiments at the Arsenal, as his laboratory notebooks show, on the production of saltpeter.

The essays on the production of saltpeter submitted for a prize to be awarded in 1778 by the Academy were disappointing and unsatisfactory. Accordingly, the Academy made no award, but announced a further competition for 1782 with the prize increased to 8,000 livres, and other sums allotted for competitors whose work merited recognition. On this second occasion the essays submitted were of a much higher standard and the prize was awarded to the brothers Thouvenel, whose essay, together with those of other competitors, was edited for publication with a historical introduction by Lavoisier. The Academy had ordered the committee appointed to judge the essays to carry out the work of editing and publishing, but almost all of this labor fell to Lavoisier who, as was usual, functioned as secretary of the committee.

Here, as in the Tax-Farm, Lavoisier had to meet and solve the problem of the recruitment and appointment of competent officials to replace the unsatisfactory staff of the previous régime, and to insure that his reforms were effected. With his fellow commissioners, he resolved to admit no candidate for appointment who did not have an adequate knowledge of chemistry, mathematics, and the construction of powder mills and refineries.

Lavoisier's interest in the manufacture of gunpowder nearly cost him his life in 1788. An experiment with a powder containing, in place of saltpeter, the new substance, potassium chlorate, recently discovered by Berthollet, was made at the latter's suggestion and by the direction of the Minister of Finance at Essonnes. It might prove a more economical product, said Lavoisier, and an advantage to the nation that first used it in war. Lavoisier had arranged with Le Tort, manager at Essonnes, and Chevraud, a Commissioner for Gunpowder, for the experiment to be made on Sunday, October 27, and he arrived at Essonnes on the previous evening accompanied by Mme. Lavoisier and Berthollet. As a precaution, the preparation of the new explosive was carried out in a small hand mill, fitted with only one pestle, in the open air, and the workmen engaged in mixing the powder were protected, in case of accident, by a stout wooden screen dividing them from the mortar in which the mixing was being done. The experiment began at six o'clock in the morning when the ingredients for sixteen livres of gunpowder were weighed out; the mixing began at seven o'clock. Although it was damped, the mixture turned lumpy and material was thrown up on the sides of the mortar. The observers, Lavoisier and Mme. Lavoisier, Berthollet, Chevraud and his sister Mlle. Chevraud, and Le Tort, with Mallet, an apprentice, and Aledin, the master gunpowdermaker, were standing on the unprotected side of the mortar opposite the workmen, who were protected by the wooden screen. Le Tort began to push down with his stick the material that was being thrown up on the sides of the mortar, but this made no improvement and so a new and moister mixture was made by adding the proper proportions of the ingredients to bring the total weight up to twenty livres. The product was still lumpy and, despite Lavoisier's warnings, Le Tort continued to knock down with a stick the lumps thrown up at each stroke of the pestle, because, he said, the powder was not yet mixed and there was no danger; and he talked jestingly to the company of the effects that an explosion might produce.

At a quarter past eight o'clock, the mixing seemed to be more advanced than had been expected and Lavoisier, after insisting that Le Tort should now cease from stirring the powder, remarked sharply on the futility of constructing screens as a protection from accidents and then standing in front instead of behind them. He gave instructions that the workmen should make only eight or ten turns of the mill at a time and then stop while the powder was being stirred, and that all present should remain behind the screen while the mill was working. This arranged, the party went to breakfast, leaving Mallet and Aledin in charge, with the promise that they would be quickly relieved. On leaving, Lavoisier gave final and emphatic instructions that operations should proceed as he had indicated. On the way to breakfast, Le Tort said that he was worried that he had left Aledin at the mill because he was a married man with a family and that it would have been preferable to put a boy there; but Lavoisier replied that, if all present stayed behind the screen, there was no danger.

A quarter of an hour later they started to return. On their way Lavoisier, Mme. Lavoisier, and Chevraud, fortunately stopped for a few minutes to show a powder mill to Berthollet, who knew nothing about the manufacture of gunpowder, while Le Tort and Mlle. Chevraud went on. Lavoisier and his party had scarcely resumed their steps, when a loud explosion was heard and a thick column of smoke rose from the scene of the experiment. When they ran forward, they found the mill completely wrecked, the mortar broken in pieces, and the pestle hurled to a distance, Le Tort and Mlle. Chevraud fatally and terribly injured and flung thirty feet away against a wall by the violence of the explosion. Mlle. Chevraud was dying and Le Tort lived about half an hour. Neither Mallet nor Aledin nor any of the workmen was injured;

they had all been saved by the screen. Had the explosion occurred half an hour earlier or three minutes later, modestly observed Lavoisier, there would have been more to regret.

Lavoisier did not feel, however, that research on the use of potassium chlorate in gunpowder should be abandoned. After reporting to the Minister the details of the fatal accident, "of which I and one who is very dear to me have nearly been the victims," he continued: "If it should be your pleasure, Monsieur, to take up a moment of the King's time with the report of this sad event and of the danger to which I have been exposed, permit me to beg you to assure His Majesty at the same time that my life belongs to him and to France, and that I am always ready to sacrifice it in his service when some advantage may follow, either by resuming the same work on the new powder, work which I consider necessary, or in any other way."

On another occasion, during the early days of the Revolution, on August 6, 1789, three weeks after the fall of the Bastille, Lavoisier, as Commissioner for Gunpowder, was in danger of another kind and again he narrowly escaped with his life. The magazines of the Arsenal were encumbered with some 10,000 livres of low-grade powder, manufactured for trading and export, and it was decided to send this to Essonnes and replace it with musket powder, which was more likely to be needed in view of the state of affairs in Paris. In fact, the low-grade powder had been on its way from the factory at Metz, where it had been made, to the ports of Rouen and Nantes for shipment. But throughout France the panic-stricken provincial officials smelled treachery everywhere in those anxious days and, patriotically insuring that the cargoes did not find their way into the hands, and the muskets, of their country's enemies, for whom they felt these consignments were surely bound, they stopped all movements of gunpowder. So the powder from Metz had been seized at Château-Thierry by order of the municipality and redirected to Paris to the Arsenal.

Here the real trouble began, after the Gunpowder Commissioners had arranged to replace this low-grade powder by musket powder from Essonnes. No munitions of war could be moved from Paris without permission of the commander of the National Guard, the Marquis de Lafayette. The Marquis de la Salle, Lafayette's chief of staff, signed the authority for the movement of the gunpowder in Lafayette's absence; and the powder was loaded at Port-St.-Paul in a boat guarded and escorted by four members of the National Guard. During the loading of the powder, the local inhabitants grew suspicious and on August 5 warned the committee sitting at the Hôtel de Ville. Bailly, astronomer and mayor of Paris, and Lafayette, unaware of the authority signed by La Salle, instead of seeking information about the movement from the Gunpowder Commission, ordered the boat to be unloaded and the powder to be remagazined in the Arsenal.

Their order made matters worse; order and counter-order produced disorder, and suspicion and rumor now ran their full course. The Gunpowder Commissioners were traitors robbing the citizens of Paris of the powder and shot that were their sole means of defense; even the escort was clapped under arrest; and a close watch was maintained on the boat. Next day there was ample protest before the assembly of the local commune. The escort was liberated and Lavoisier explained why the movement had been planned; but the assembly, with the intention of calming the public disquiet, sent two of its members as representatives to conduct an inquiry at Port-St.-Paul, where the boat was now surrounded by a mob of guards and citizens from the surrounding districts. The two representatives ordered the powder to be taken back to the Arsenal and then proceeded with their examination. Four barrels, chosen at random, were opened in the presence of Lavoisier and one of his colleagues in the Gunpowder Commission; the contents were recognized as low-grade trading powder. The necessary documents were now signed and all seemed in order, except for the surging mob at the gates of the Arsenal demanding the arrest of Lavoisier and his colleague, both of whom were now led to the Hôtel de Ville, followed by the noisy and angry mob. Here Lavoisier coolly gave a clear explanation about the gunpowder and had no difficulty in vindicating his actions; and so he and his colleague were set at liberty.

But, outside, on the Place de Grève, passions were rising. The mob, still crying that there must be treachery somewhere, heard that it was La Salle who had signed the order for the movement and promptly turned their vengeance on him and pushed their way into the Hôtel de Ville. Lafayette was helpless, as the mob, deaf to his appeals and chidings, ranged everywhere, searching for La Salle, even in the clock tower, and calling for his death. To satisfy them Lafayette ordered a party of soldiers to find him so that he could defend his action. La Salle, warned in time, escaped, but was later arrested and spent some weeks in prison until liberated by the National Assembly on September 5. Bailly was at Versailles with his wife and did not return to Paris until half-past eleven that night, when he first heard of the disorder and was told that Lavoisier and Mme. Lavoisier had been arrested. Accompanied by Mme. Bailly, who wished to share the danger with her husband and to help to save their friends, he hurried to the Hôtel de Ville, but when he arrived there at midnight order had been established. Lafayette had cleared the Hôtel de Ville and dispersed the mob on the Place de Grève. Two days later 10,000 pounds of musket powder from Essonnes were magazined in the Arsenal and public clamor died down; but, among the unruly, suspicion was unallayed and the incident was not forgotten.

XVII

REPORTS TO THE ACADEMY

DURING HIS MEMBERSHIP OF THE ACADEMY, LAVOISIER TOOK a leading part in all its activities. Apart from original memoirs, more than fifty in number, in which he presented for publication the details of his great discoveries, he compiled either alone or conjointly over two hundred reports, large and small, on an extraordinarily wide range of subjects. Of some of the more striking and important of these, which reveal Lavoisier's humanity as well as his scientific temper, we shall now give some account.

Prisons

INVITED IN 1780 BY MINISTER NECKER TO REPORT WITH other members of the Academy on the prisons of Paris, the Châtelet, the Conciergerie, and For-l'Évèque, Lavoisier recommended reconstruction of the old buildings and also the erection of new ones on a different and more humane plan. Mme. Necker was much interested in this social problem; it was to her that the report was presented and it was to her that Lavoisier complained when prison officials, opposed to reform, were obstructing him and his colleagues in their work. "Would you believe, Madame," he wrote on March 25, 1780, "that, while the prisons are in a sense open to all, we found it hard to get inside them? Our mission seemed at first to have given offence to the magistrates. We had to be very tactful and take many precautions so that our co-operation in a task so worthy of a good administration should not be taken amiss; happily we succeeded in conciliating all opposition and, after first meeting some obstruction, later we had nothing but praise for the care and consideration of the magistrates and for the orders that they made that nothing should be hidden from us. We felt that we were carrying out your wishes in taking precautions that were perhaps excessive that our mission should give offence to none."

Conditions in the French prisons had become a matter of public disquiet and a source of embarrassment to the Government; they were as bad in other countries, except in Holland and Germany, where prisoners were properly accommodated and fed and given useful work to do. The grim facts were all set down by Lavoisier. The prisons of France were overcrowded almost beyond credible limits, filthy, verminous, hotbeds of evil and disease and corruption. They housed alike criminals, civil offenders, and military prisoners, and also those awaiting trial, among whom were, of course, the innocent as well as the guilty. Dark and damp, foul and unventilated, they stank and festered, polluting all who entered them. To describe them as insanitary would be facetious.

Lavoisier inspired the whole report drawn up in association with his colleagues, and it is entirely in his handwriting. It included many recommendations and described in detail how these should be carried out. Where water was necessary, sources and means were indicated; drains and sewers were specified, not merely advised in general terms; means and methods of ventilation were named; cleansing of the buildings, the prisoners, and their clothes, was similarly put down in detail. Adequate supplies of water were essential; fresh air must be conveyed into the cells and circulated through the prisons. Each prisoner must have enough room to sleep; space for this purpose was specified. Cells must be regularly cleansed, drains and sewers regularly flushed. Civil offenders and military prisoners must not be put with criminals, who should be kept apart. All those detained should do some useful work and there should be yards to enable them to take regular exercise. The prisoners should be given clothing and should have some necessary furniture in their cells as a recognition that they were human beings, not animals. All buying from prison officials should cease. It should be realized by those responsible that the prisoners were subjects of France who would once again go out into the world, against which time it was the duty of all those charged with their care to insure that they did not re-enter it spreading contagious disease about them in towns and villages, ships and colonies.

Lavoisier, the report clearly shows, had studied closely what had been done in other countries. He was deeply impressed by the work of John Howard in England and he referred to statements about gaol fever by Lord Bacon, Sir John Pringle, and Dr. Lind. Unfortunately, Necker was again dismissed in 1781 and the recommendations of the report were not put into effect.

The "aerostatic machine" or balloon

THE "AEROSTATIC MACHINE," OR BALLOON, WAS ANOTHER problem upon which Lavoisier's advice was sought by the Government. On June 5, 1783, at Annonay the brothers Montgolfier had sent up to a considerable height a balloon, consisting of a hollow globe made of linen and paper with a diameter of 35 feet, weighing 450 pounds, filled with hot air, and carrying 400 pounds. The balloon travelled about 2,500 yards before descending. When d'Ormesson, the Controller-General, called for a report from the Academy of Sciences, it responded by appointing commissioners for this purpose. Lavoisier was named as one of the investigators, and again it was he who seems to have played the leading part in the work. Experiment was resolved upon and the younger Montgolfier was invited to make a balloon at the expense of the Academy. The Government, however, considered that the invention was so important and the expense likely to prove so much beyond the Academy's resources that all costs should fall to the State.

The report opened with brief historical reference to writers on the problem of flight, such as Roger Bacon, Lana, Hooke, Borelli, and Gallien. The elder Montgolfier, it added, had succeeded in November, 1782, in sending up a small balloon filled with hot air to a height of 70 feet. Shortly afterwards, he and his brother made a similar successful experiment with a larger balloon, 650 cubic feet in volume, which reached a height estimated at 600 to 900 feet. Then at Annonay they made the experiment on which the Academy was invited to report by the Government.

The commissioners then described the experiments made under their direction. After some mishaps, on September 18, 1783, a balloon, of nearly 37,500 cubic feet capacity and weighing about 800 pounds, filled with hot air from burning straw, ascended successfully into the air for 5 or 6 minutes, descending as the hot air cooled. An experiment before the King and Queen and the Court had been fixed for the next day at Versailles and was most successful. The balloon rose to a height of more than 1,400 feet, although carrying an extra weight of 200 pounds, and travelled nearly 2 miles before descending gently 10 minutes later: its passengers, a sheep, a cock, and a duck, first of all travellers by air, were none the worse for their voyage.

Further experiments were considered necessary, especially to determine whether a balloon could be made large enough to carry passengers and how its flight could be controlled. A large balloon was made, 45 feet in diameter and 70 feet high. It weighed about 1,400 to 1,500 pounds, and carried a small stove for fuel to produce the hot air and a wicker cage for the balloonist. In this apparatus on October 15, Pilatre de Rozier made the first human ascent in a captive balloon. He rose to a height of about 100 feet, the balloon being anchored by ropes. By opening or closing a valve in the balloon, he could release or retain the hot air supplied by the stove and accordingly descend or ascend at will. Similar successes followed and Giroud de Villette and the Marquis d'Arlandes also ascended in this way to even greater heights. Then on November 25, 1783, the first passenger flight in a balloon was made by Pilatre de Rozier and the Marquis d'Arlandes before the Dauphin and the Court. They rose to a height of more than 2,000 feet from the Jardins de la Muette, crossed the Seine, and descended near the Fontainebleau road. They were about 17 minutes in the air and had travelled nearly 5 miles in man's first aerial voyage.

In describing this accomplishment the commissioners referred also to the successful flight on December 1, 1783, of Charles and Robert in a balloon filled with inflammable air. They pointed out that such a balloon did not need constant recharging in flight as did the hot-air balloon, although the latter was more conveniently and easily filled. The answer to the question of what gas to use must wait for further experiments.

The uses and applications of the invention were, however, said the commissioners, so many that a volume would be needed to discuss them and they would increase if the direction of the flight of the balloon could be controlled. They recommended that the Academy award the Montgolfiers the prize of 600 livres, left by an anonymous benefactor for inventions in the arts, as discoverers whose work would mark an epoch in the history of human invention. The report was submitted to the Academy on December 23, 1783, and adopted unanimously.

Further work was done on the problem by the commissioners in December, 1783, and January, 1784. At their meetings Lavoisier set out as the desiderata for perfecting the balloon a light strong envelope impermeable to air—and especially to hydrogen—even at low pressures, a light cheap gas easily obtainable everywhere at any time, a method of controlling the ascent and descent of the balloon over a range of 1,200 to 1,800 feet, and a simple means of steering it in a given direction. Since those modest beginnings the world has seen the development of the airship. When technical research on that ill-fated craft ceased, the problems remained essentially as Lavoisier had first stated them. Flight has been achieved by other means.

Mesmerism

MUCH OF LAVOISIER'S WORK IN THE ACADEMY TOOK THE form of reports made at the request of the Government. A very different problem arose in 1784, and official action became necessary. Paris, and Europe also, had been stirred for several years by the cures said to have been effected by Friedrich Anton Mesmer (1734-1815), an Austrian physician, who had first come to Paris in 1778. Asserting that the human body was endowed with "animal magnetism," he claimed that disease could be cured by the passage of this "magnetism" from the healthy to the sick, by touch and other means, and even from a distance. "Mesmerism" became so popular in Paris that the Government asked for the opinion of the Academy, since the popularity of the new treatment as opposed to orthodox medicine had become a matter of urgent public importance.

The Academy nominated as commissioners Lavoisier and some of his colleagues, including Franklin and Bailly. As on many other occasions of this kind, the report was drawn up by Lavoisier and is in his own handwriting. It outlined the plan to be adopted, after a critical study of the books on mesmerism, and had, by that study, given in advance the general form of the conclusions at which the commissioners should arrive. In its opening section, it pointed to the great obstacle in medical experimentation, namely, the difficulty of establishing the cause of a cure:

> The art of drawing conclusions from experiments and observations consists in evaluating probabilities and in estimating whether they are sufficiently great or numerous enough to constitute proofs. This kind of calculation is more complicated and more difficult than

> it is commonly thought to be; it demands a high degree of sagacity and is for the most part beyond the powers of the generality of mankind. It is on their errors in calculations of this kind that the successes of charlatans, sorcerers, and alchemists, are based, and so formerly were those of magicians, enchanters, and all those in general who deceived themselves or who sought to take advantage of a credulous public. In medicine, especially, the difficulty of evaluating probabilities is greater. As the principle of life in animals is an ever active force unceasingly tending to overcome all obstacles, so Nature, left to her own devices, cures a great many diseases; and when remedies have been applied, it is extremely difficult to decide how much is due to Nature and how much to the remedy. So, while the multitude considers the cure of the disease to be a proof of the efficacy of the remedy, the only inference in the eyes of the judicious is a degree of probability, more or less great, and this probability can be turned into certainty only by a great number of instances of the same kind.

Having made very clear in this passage the necessity for medical skepticism and the danger inherent in drawing conclusions from medical experiments, the report went on to say that the commissioners, not being physicians, considered that their duty appeared to them to consist in deciding whether or not there were such an agent as "animal magnetism." It was not to establish the correctness or otherwise of the claims to have effected cures, since that lay in the province of the physician and since, in any event, cure often depended on many factors and could be established as a general conclusion only by the study of a very large number of cases.

Mesmer was absent from Paris, but his disciple, Deslon,

was conducting a considerable practice and was ready to give the commissioners every possible assistance in their inquiries. Deslon's treatment was observed. Patients sat round an ebony chest, each gripping a movable iron handle which could be applied by means of a hinge to the part of the body affected by disease. A cord, passed round their bodies, joined them one to another. They linked hands, each pressing the thumb of his neighbor. A piano played and sometimes there was singing. Deslon carried a wooden wand, fitted with pieces of iron, to conduct the "magnetism" during the proceedings, and passed it before the eyes as well as before and behind the bodies of his patients. He also practised cure by touch, applying his hands to the part of the body that was said to be diseased.

What of the results? To some, nothing happened: they remained calm. Others coughed and spat, felt pain, grew hot and sweated, or even experienced crises, in which they suffered convulsions in all their limbs, or almost choked or cried or yelled out or hiccoughed or burst into peals of laughter. Gradually, the crises passed off.

The commissioners failed to detect the alleged "animal magnetism" by every known scientific test. It was neither magnetism nor electricity. Deslon, however, pointed to the "cures" that he had achieved and insisted that the commissioners should limit their examination to such evidence. On the contrary, they decided against such a one-sided interpretation of their duty and resolved to investigate the question of the reality of the effects ascribed to "animal magnetism." Experimenting on themselves by Deslon's method and with his apparatus, they experienced no crises or anything in the slightest degree comparable, whereas, they pointed out, a real cause should always produce the same effect in the same conditions. They therefore suspected that perhaps imagination had much to do with these crises.

Further experiments were made at Franklin's house at Passy and at Jumelin's house in Paris, all details being carefully planned beforehand, and they confirmed their suspicions. They were able to produce the crises by telling the patients, who were previously blindfolded, that they were being "magnetized," when, in fact, they were not being so treated. Then, the crises over, the patients, still blindfolded, were "magnetized," and no crises occurred.

At Passy an interesting experiment was made with two of Deslon's patients, who were said to be very susceptible to "magnetism." Both experienced crises, while blindfolded, under the belief that Deslon was in the room and had been instructed to "magnetize" them. In another test, Deslon was invited to "magnetize," according to his practice, one tree in a row of trees in an orchard, while one of his patients, a young man who was said to be very susceptible to "magnetism," was kept out of sight. The blindfolded patient was then led to different trees at a distance from the one that had been "magnetized." But, at the very first tree, he felt some effects which gradually increased as he passed along the row, and at the fourth, although it was "unmagnetized" and far from the one that had been so treated, he collapsed in a crisis, his limbs became rigid, and he lost consciousness.

In another experiment Deslon was asked to "magnetize" one of a dozen china cups, this cup being previously marked with a sign known only to the commissioners. The cups were presented one after another to one of the especially sensitive patients, the "magnetized" cup being kept well back towards the end of the series. But the fourth cup produced a crisis and, when the patient asked for a drink, he was given it in the "magnetized" cup. The crisis passed and the patient announced himself relieved. The cup and the "magnetism" had therefore produced effects contrary to their alleged properties; for the crisis had been produced in the absence of the so-called "magnetism" and allayed by its presence. Some time later, the patient had a further crisis after being led into another room. Hearing the commissioners busy on other experiments and knowing that the purpose of his visit was to undergo tests, he thought he was being "magnetized" and fell into a crisis. Imagination, in all these cases, could produce the effects without Mesmer's "magnetism."

It remained to be proved that the "animal magnetism" could not produce the effects without imagination; and a corresponding series of tests was successfully made, in which patients said to be especially sensitive were "magnetized" according to Deslon's methods without their knowledge and experienced no crises or similar effects.

All was, therefore, due to imagination. The practice of mesmerism was not only exposed, but also condemned by the commissioners as likely to have harmful results on those who subjected themselves to it.

The hospitals of Paris

IN 1787 THERE CAME ANOTHER REQUEST FROM THE GOVERNMENT to the Academy of Sciences; it was for the appointment of commissioners to report on the hospitals of Paris. Again Lavoisier was nominated to serve. The commission considered first the old problem of removing the ancient hospital, the Hôtel-Dieu, to a new site.

Conditions there were almost as bad as those in the prisons. There was not room for the sick of a population as great as that of Paris. The building was unhealthy; patients were crowded, many in the same bed, where they could neither rest nor sleep; fresh air was unknown. Lunatics were admitted to the hospital where their cries disturbed the sick; patients with contagious diseases mixed with those suffering from other complaints; smallpox and fever cases were in the same ward; noise was everywhere. The appalling mortality exceeded that of any other hospital in Europe. In the St.-Paul ward, there were 111 beds and 272 patients. Through the ward was a passage to the stores of bread and wine, and to the cellars for linen and sand and for the removal of soiled linen. Wood, food, and clean linen, were carried in by this route; and the ward was a waiting room every afternoon for the poor who wished to consult the surgeon.

Such were the conditions reported by the commissioners. How could a patient, they asked in effect, recover in such surroundings? No reform could be more urgent. They inspected other hospitals and found some of them to compare favorably with the Hôtel-Dieu, although they noted defects. Within a year, their detailed report was complete, together with two further reports with plans for four new hospitals; every detail was considered and mortality rates from many European hospitals were quoted. No point escaped attention in the defects of the old hospitals and in the plans for the new ones in a long report that ran to forty thousand words and was marked by every humane consideration for the sick poor.

These reports are a few among many such public tasks carried out by Lavoisier as the most active member of a number of commissions instituted at the request of the Government. Throughout, his work was marked by a broad humanity towards his fellow citizens, whom he always regarded as human beings, as men and women and children, and subjects of France.

XVIII

THE FIRST EXPERIMENTAL FARM: FRÉCHINES

LAVOISIER'S FAMILY, STEADILY RISING IN THE SOCIAL SCALE during two centuries, had originated in the sturdy peasant class that has given France so many of her greatest sons and still forms a half of her people. The peasant, said Professor Denis Saurat in the grim days of 1940, is "indeed the body of France;" and "no tyrannical system of government can ever be built on such people . . . the spirit of the peasant is an essential part of the spirit of France and endures in all the achievements of the nation."

In his vacations from the Academy Lavoisier maintained his family's connection with the countryside. He had inherited a house at Le Bourget and in 1778 he bought the estate and château of Fréchines (Plates 19 and 20), between Blois and Vendôme. Already in his geological surveys with Guettard he had seen much of the miserable condition of French agriculture and had become inspired with the idea of improving that most vital of all industries.

19. The estate of Fréchines and its neighbourhood from Cassini's map of France. Fréchines, in the parish of Ville-Francoeur and Champigny, is half-way between Blois and Vendôme, the road connecting which is shown traversing the map from S.E. to N.W. Lavoisier acquired also the lands of Champ-Renault, immediately north of Fréchines, and later other lands at Chapelle-Vendômoise; and he also bought the house and lands of Thoisy, shown on the map immediately south of Fréchines, but sold this house in 1793 to his fellow-Academician, Charles Augustin Coulomb (1736–1806), the physicist who demonstrated that the force between magnetic or electric charges followed an inverse-square law. Fréchines and the other lands were the scene of Lavoisier's experiments in agriculture; and it was here that he introduced the cultivation of the potato into Loir-et-Cher.

At Fréchines he made many experiments for this purpose, and in 1788 he reported them to the Society of Agriculture in Paris in a memoir entitled "Results of some agricultural experiments and reflections on their relation to political economy." This work is of historic importance both in agriculture and in economics. It shows Lavoisier

20. The Château of Fréchines.

skillfully applying to practical problems in farming the same masterly scientific analysis that marked all his chemical researches.

Opening the account of his experiments, Lavoisier explained to the Society of Agriculture that for ten years he had worked to collect materials for a treatise on agriculture. He was as yet, however, far from achieving his aim and from presenting decisive results, but was at least able to survey what he had so far carried out. At Fréchines, when he began his work, the average yield of wheat had been only about five times the seed and, while this appeared to him to be due to some extent to the poor quality of the soil, he felt that it was due still more to bad cultivation and to lack of money and means. Most of the farmers in the district had only four or five cows and eighty sheep on a farm with three ploughs. They did not make or cultivate meadows and so had no fodder for cattle during the winter. They knew nothing of the practice of folding sheep, in which parts of a field were successively fenced off to be occupied, if already ploughed, or grazed in turn by the sheep so that the land was manured and afterwards yielded increased crops. They spread only two or three loads of manure per acre on their land each year. The condition and practice of farming in the district were bad and Lavoisier resolved to help the farmers by setting them an example of good husbandry directed by sound principles.

As a beginning to his enterprise, Lavoisier decided to cultivate on his own account a part of the estate that best suited his purpose where the soil was poorest. For comparison, he took half-shares in arable land on three other farms; and further to this purpose and to determine yields on a wider scale, he leased a farm from another landowner. Thus he farmed 240 acres on his own and had a half share in 600 acres, as well as a farm on lease.

He had already observed that the low fertility of the land in his district was mainly due to lack of manure, which in turn was due to lack of stock. To those who had given no thought to this problem, Lavoisier pointed out, nothing would seem easier than to revive a languishing agriculture. Only stock and money were needed. But before stocking a farm with cattle and sheep, the farmer must be able to feed them. His first problem was fodder. Moreever, beasts did not provide more manure even when well fed in stables; for straw was necessary, since a satisfactory fertilizer for the land was made only by the mixture of animal and vegetable matter.

The first practical step towards improvement was, therefore, the preparation and cultivation of meadows and, generally, the increasing of fodder for stock. At Fréchines, this was an entirely new practice and Lavoisier had no experience to guide him. He could learn nothing from the example of his neighbors about what crop best suited the soil or what kind of cultivation was best adapted to the different crops. He realized that he would have to make many experiments and study his land before he could indulge in any general speculations. It was only at the end of three years that he recognized that lucerne did not succeed on his soil and that it was almost impossible to protect it from the plant parasite, cuscuta or dodder; that his land was more suitable for sainfoin; that clover succeeded in rainy years, but often did not germinate in dry years; and that with care by suitably preparing the land he could grow turnips and potatoes in the open fields, vetches and peas on the fallow.

With these facts won from the experience of the first years, Lavoisier began to broach the plan that he had formed. From the beginning he introduced the folding of sheep, despite much local prejudice which only time and experience would conquer. Four to five hundred sheep, folded from midsummer to the end of October, manured about 25 acres without consuming straw. This first experiment enabled him to spread on 55 acres the manure that he had previously had to spread on 80 acres, and to spread it at 3 loads instead of 2 per acre.

As it was only by increasing manure that the farmer could increase straw and only by increasing straw that he could increase manure, this twofold aim could be attained only by a slowly progressive advance. Lavoisier, however, accelerated the process by buying straw and by putting into his own farm the straw from the farm that he had leased. The straw, consumed by the cattle that his meadows had allowed him to feed, gradually increased the amount of manure. Steadily during seven or eight years he was able to increase the distribution of manure on his lands from 2 loads per acre to 6 or 7, and he hoped in another few years to raise it to 10 loads per acre or perhaps more. His yield of straw had already almost doubled, but what he found very remarkable was that the yield of wheat had scarcely increased at all or, at least, that it had increased negligibly in comparison with that of the straw.

At the time of drawing up this first account of his work, August, 1787, he had about 25 acres of meadow in good condition, 2 acres of turnips, 1 acre of mangolds, 1½ acres of potatoes, some clover and also vetches on the fallow, a herd of 20 cows shortly to be raised to 30, a flock of 500 sheep folded, and the wherewithal to feed 300 sheep during the next winter. His barns and his granaries were no longer sufficient to store the abundance of his fodder; and his harvest of oats already sensibly exceeded his consumption.

Thus, as a result of eight years of cultivation, Lavoisier had notable successes to report. He had obtained a considerable increase of fodder and an abundance of straw and manure, but little increase in the yield of wheat or in his proceeds in money. Advances in agriculture, he pointed out, were very slow. But he recognized with some difficulty and learned to his cost that, no matter what attention was given to these problems and no matter what economy was brought to bear upon them, it was impossible to reap a profit of 5 per cent on the capital invested. This, he concluded, must doubtless be the reason why well-off farmers, living near Paris, who were able to save money, preferred to invest it in public loans and funds rather than to use it for the improvement of agriculture, rather than to "plough it back into the land" in our modern phrase. This in turn was clear proof that the needs of the Government were maintaining the interest on money at too high a rate in France. Here was an insurmountable obstacle to the improvement of agriculture, and probably of other industries; and he urged that it was vital to national prosperity for the Government to consider means of reducing it.

Another conclusion from these results, observed Lavoisier, and one that was likewise a result of the slow progress even in a good system of husbandry, was that, in the present state of affairs, French agriculture could be improved and revived only by wealthy landowners who would be willing to invest some of their money in the cultivation of their land, or by the better-off farmers who could be regarded as landowners if they had very long leases, for example, leases of twenty-seven years. Average farmers were far from being able to make the advances necessary to stock a farm and, when they were in a position to do so, they found that their money could be invested more profitably in Paris and other cities. Moreover, it was only after eight or ten years of expensive cultivation that the effects of the improvements made in farming began to appear and this period exceeded the term of most leases.

Lavoisier had had to face one very serious difficulty which would have upset his plans entirely and perhaps irremediably, if he had not fought it with all his resourceful energy. This was the drought of 1785. Bad for most of the kingdom, it was still worse for land such as his, which tended to harden and crack. Most of the methods he tried were recommended to him by his fellow members of the Society of Agriculture. He sowed vetches on the fallow in the months of May and June, and even July, and cut the crop green in the course of September; and he obtained a sufficient amount of fodder. He sowed buckwheat immediately after harvesting his rye; this was, indeed, not a complete success, but it gave him a fodder for the winter which the cattle ate for lack of another. He found a more effective remedy in the cultivation of turnips. Those sown in July and the beginning of August matured and, although they were of no great size, because the soil had not been prepared for the crop, they provided ample feeding for the stock. Those sown still later supplied green fodder for the autumn and the following spring. During this year of drought, Lavoisier's flocks and herds suffered little. He had no cows or sheep sick and he escaped with only some decrease in the yield of milk and butter. An interesting detail, which he noted, was that this decrease continued to make itself felt long after abundance had been re-established and it was not until a year later that the cows again gave their usual quantity of milk.

Lavoisier soon realized how difficult it was to follow from Paris the prosecution of an enterprise on such a scale, and to direct intricate and difficult experiments from a distance of forty leagues. He therefore resorted to a plan and obtained assistance in this work, which he now described to the Society.

One of the chief difficulties in agricultural experiments, he explained, was to know exactly the areas of the various plots of land cultivated. Mistakes that might be made regarding this changed all the proportions and would exclude the possibility of comparisons. Impressed with the importance of this problem, Lavoisier began by making accurate maps and plans in duplicate for his own farm and the farm that he had leased. These showed the land in every detail and all its divisions and subdivisions as circumstances demanded. One set of these maps and plans was kept at Fréchines and the duplicate in Paris.

Another difficulty was to follow a large number of experiments at the same time without confusion. So at Paris Lavoisier kept registers in which each plot of land had its section or "chapter," recording its cultivation during nine years and the weights in detail of all the crops that it had successively yielded. By combining with these registers an alphabetical index of crops and cultivation, he was always able to get at the facts that he needed.

A third difficulty, he added, was to determine exactly the amounts of the crops obtained. Careful management and close attention were needed at the time of harvest and threshing. He considered that this part of his work was perhaps the most satisfactory. In experiments that demanded accuracy, all the sheaves or trusses, according to the kind of crop, were weighed as soon as they were tied. For ordinary crops, it was enough to weigh eight or ten by load as they reached the barn, and the weight of the rest was calculated by the number of sheaves or trusses multiplied by the average weight. By this means all went into the barn by number and weight. The same care was taken when threshing took place, and all came out also from the barn by number and weight. Thus in the threshing of wheat, for example, the wheat, the straw, the chaff, and so on, which came from the threshing, were weighed separately, and thus Lavoisier obtained not only the total weight of the crop for each plot of land and its subdivisions, but also the weight of each kind of product. Finally, as the plots were of unequal size and it would be otherwise impossible to compare one with another, he converted all the results by calculation to what would have been obtained if the area of each plot had been 1,000 square toises in area (about 4/5 of an English acre).

It was not often that Lavoisier was able to visit Fréchines more than three times a year and he could not stay for more than a fortnight or three weeks. So far as he could, he chose the times of autumn sowing, spring sowing, and harvest. In his absence, his experiments were under the observation of a neighbor and friend, whose zeal, activity, and intelligence, in agricultural matters were well known to the Society.

All that Lavoisier had so far reported applied only to his own farm. The leases for half-shares that he made with his farmers obliged them to maintain a certain area of meadow land and offered a bounty for the number of cows or sheep above a stated proportion, especially for folded sheep. But, although he made advances in stock, and left nothing undone to rouse the interest of the farmers in increasing their head of cattle, he was sorely disappointed to find that his bounties were not earned, and that he was obstructed in this direction by an attitude of mind more difficult to overcome than most physical obstacles. The *taille,* or tax on property, under the system then prevailing in France, Lavoisier pointed out, increased in proportion to the circumstances of the farmer, which were estimated only by the number of his horses and cattle, so that those who were best stocked with farm animals paid the most tax. The good farmer was penalized. The whole purpose of Lavoisier's premiums for the promotion of agriculture was therefore defeated by what were, in effect, repressive premiums that worked in the opposite direction, and his plans were in a sense destroyed before they had taken shape. He hoped, however, that the good sense of the Provincial Assemblies of France would be applied to this urgent problem and that the principles of taxation that they would adopt would prevent mistakes such as these, which were sufficient in themselves to hinder all progress in agriculture.

Turning to the farms in which he had a half-share, Lavoisier observed that, although they did not present the same difficulties as his own farm and although he could not promise himself the same success, it might be useful for him to describe the details of the method that he had adopted to ascertain the yield of their harvests. It was stipulated in his leases that at the time of harvest the sheaves should be ranged in two rows, each of which contained an equal number. His overseer chose the row that he thought fit and had it loaded and put away on Lavoisier's account in a barn kept for his own use. The sheaves were counted and weighed when they were put in the barn, the number and weight of the sheaves from each plot being separately recorded. For these farms he had maps and plans and kept registers at Paris as he did for his own farm, so that he could keep himself informed of all the details of these long experiments.

So far, he explained to the Society, he had described the work on which he had been engaged only from the general point of view of improving and restoring agriculture. There was, however, another object which he regarded as more important, namely, to provide the economists with definite information about the distribution of territorial wealth. He had tried to determine exactly for all the land under his cultivation

(1) the number of sheaves that went to the owner of the tithe,
(2) the number necessary to pay for the labor of harvest,
(3) the number required to pay for the labor of threshing,

(4) the number reserved for next season's sowings,
(5) the number used by the farmer to cover all expenses, such as the upkeep of vehicles, harness and equipment, replacement of stock, etc.,
(6) the number necessary for food and maintenance of the farmer and his family,
(7) the number required to pay the rent to the owner of the land, and
(8) the number that fell to the revenue for payment of the land-tax and of the various taxes on commodities.

Lavoisier stated that he would not refer to the interest on the first advances made by the farmer, because most of this was included in the expenses of maintaining vehicles and equipment and restocking with cattle. This way of considering territorial wealth in its own kind, without the complication of converting the value into money, in his opinion, greatly simplified the problem; and he felt that the treatise that he had in mind on this subject would throw a new light on political economy.

However, as a first conclusion, it already appeared that the landowner, at least in the district in which Lavoisier farmed, took away between a quarter and a third of the harvest and that the taxes carried off nearly as much. After the cost of seed had been deducted, about a third was left to the farmer for his maintenance, his food, his expenses in farming, the return of interest on his capital, and all his other charges.

Yet, added Lavoisier, the most distressing part of this picture was that, from the profit of a declining agriculture, such as that of the greatest part of the provinces of France, there remained almost nothing at the end of the year for the unfortunate farmer, who thought himself lucky when he could lead a miserable and stunted life, and when in years of plenty he could save some money that was before long absorbed in average and unfruitful seasons. And the picture ought to distress rather than surprise. It was quite plain that, in the constant struggle of interests between the landowners and the farmers, the landowners raised the price of leases to the highest possible figure and thereby encroached to the utmost upon the farmer's means and reduced him to a narrower subsistence.

The condition of the farmers was not nearly so bad in countries where agriculture was prosperous and progressive, and especially where the system of long leases had been introduced. There the landowner could never let his land except on its past yields; and the farmer, from the moment when he signed the lease, became owner of all the progressive improvements that he could make during a period of twenty-seven years. Such a reform was an important aim in England and even in Flanders, and it was sorely needed almost throughout France. The French farmer had only one means of protecting himself, or at least of indemnifying himself, and that was to increase the price of his produce, but his attempts in this direction were limited by the opposing interests of the consumers.

Lavoisier emphasized that the account that he had given of his work was necessarily abridged and that he did not wish to encroach upon the larger treatise that he had in view. What he had reported, he considered, would suffice to show that political economy must be studied, not only theoretically, but also practically, by farming on a large scale for many years. His experiments had already cost him nine years of care and toil and much money, for which he could not hope to be indemnified. They could be completed only if he followed the same plan for several years to come. He had, however, learned some important principles which even the most informed persons had glimpsed but vaguely. He hoped that one day he would be able to contribute to the prosperity of the nation by stimulating public opinion through his writings and his example, and by inducing the great landlords, the capitalists, and the comfortably-off, to put their available money into the cultivation of the land.

"Such an investment of money," he concluded, "does not offer, it is true, glittering speculations on the stock-market or on the fluctuations of public loans, but it is not charged with the same risks and the same reverses; the successes that are gained draw tears from none; they are, on the contrary, accompanied by the blessings of the poor. A wealthy landowner cannot cultivate and improve his farm without spreading comfort and happiness around him; rich and abundant crops, a numerous population, and a prosperous countryside, are the rewards of his exertions."

The records of these long experiments made by Lavoisier and his discussion of them reveal how he carried over to his agricultural studies the skilled analysis that had characterized his scientific work, and how he broke new ground both in applied science and in economics. The memoir, much of which we have translated here almost in the author's own words, marks a historic development in the study of the oldest and most important human industry.

With regard to Lavoisier's energy and zest for work, we may remind ourselves that these busy years saw also the development of his theory of combustion, his experiments on the composition and decomposition of water, his criticism of the phlogiston theory, numerous memoirs submitted to the Academy, and his work, to be discussed presently, on a new chemical nomenclature, together with heavy responsibilities in the Academy, the Tax-Farm, and the Gunpowder Commission. And yet he had found time to plan and operate the first experimental farm in history, forerunner of the national agricultural research stations of all the great nations, and even today largely unknown outside his own country.

We get one last glimpse of Fréchines. In 1793 a manuscript note by Mme. Lavoisier stated that the yield of wheat had been doubled and was more than ten times the seed from which it had been grown, while the head of cattle had increased fivefold. Later in that year, when Lavoisier's property was "sealed" by order of the Revolutionary authorities from Paris, the stock included 8 farm-horses, 29 cows, 270 sheep, and 8 pigs, the total revenue of the farm and the other land that he was working being estimated at from 25,000 to 30,000 livres.

In 1803, nine years after Lavoisier's death, Mme. Lavoisier sold the château and estate of Fréchines and later the house and the lands passed to separate owners. Today the château has become a *maison de repos* in the charge of Catholic sisters from Blois and externally it is unchanged. Beneath the fine old trees, including some magnificent cedars, Lavoisier must have walked during his visits there; to the south, towards Blois, is a delightful tree-lined vista. A short distance away is the little village of Ville-francoeur, for the children of which Lavoisier built the first school, long since demolished. In the church is a beautifully decorated statue of St. Laurent, so clearly of the eighteenth century that it is not unreasonable to hazard a guess that it may have been the gift of Lavoisier. Further off are the various other lands that he farmed. This old house will always be interesting as the only habitation of Lavoisier that is now standing. In the grounds there is the earlier château, built in the fifteenth century and still inhabited, where, according to tradition, Catherine de Medici once lived.

XIX

AGRICULTURE AND ECONOMICS

WHILE THE "PHILOSOPHES" INSPIRED THEIR COUNTRYMEN, and indeed all Europe, not so much with the necessity of democratic government, but more emphatically with the ideal of liberty in all its forms as the vehicle of ordered and reasoned human progress, the economists of France applied their criticism to the excessive State regulation of their country's trade and saw that a similar freedom was becoming daily more necessary.

Under Quesnay and later Turgot, as we have seen, there arose the school of economists known as the physiocrats, who considered that the land was the only source of a nation's wealth and that freedom in trade and commerce was essential to national prosperity. Alert and interested in economic problems, Lavoisier adopted the views of the physiocrats with Turgot, Dupont de Nemours, Malesherbes, and others. While he admitted the justice of Quesnay's views on the freedom of trade, he thought that the land was the most important, but not the only, source of national wealth. He held that, while agriculture, first and most important of industries, must aim at developing that source to its fullest extent, other industries which increased by their labor the value of what the land produced were likewise important to the nation. He felt that there was excessive State regulation and a too complex system of taxation, as he had seen in his work as a Farmer-General of Taxes; and he wished to free the farmer, the manufacturer, and the trader, from the economic fetters that prevented them from the full development of their several contributions to the prosperity of the nation. He had long advocated a uniform system of taxation for, and free trade between, the provinces of France. His experiments at Fréchines had pointed the way to an improved agriculture and had revealed the defects and the economic burdens of the current system.

The Royal Society of Agriculture and the Committee of Agriculture

THE SOCIETY OF AGRICULTURE WAS FOUNDED IN PARIS IN 1761 and, although it had many distinguished and able members, waxed and waned in its activities from time to time. Lavoisier was elected a member in 1783. After a reorganization, its members began to publish volumes of memoirs in 1785 and the Society gained so much renown that in 1788 the King conferred upon it the title of Royal Society of Agriculture. Its existence and its work alike were reflections of the politics and economics debated in eighteenth-century France. Its correspondents included Arthur Young, Sir Joseph Banks, Franklin, Washington, the Duke of Parma, and other eminent foreigners.

More important, however, was the Committee of Agriculture, instituted by Calonne, when he was Minister in

1785, as a consultative body to give opinions to the Controller-General of Finance on memoirs submitted to him, and to advise him of those matters to which the Government should give their attention. It was linked with the department concerned with taxes, the commissioner of which, Gravier de Vergennes, presided over its deliberations. Its members were d'Arcet, Tillet, Poissonnier, and Lavoisier, all of the Academy of Sciences, Dupont de Nemours, and Lubert, first secretary in the department controlled by Vergennes, whose jurisdiction included the *taille,* the societies for agriculture throughout France, drainage and clearing of land, and so on. Dupont de Nemours had devoted his studies entirely to economics and agriculture. The four Academicians included the chemist, d'Arcet, appointed director of the porcelain factory at Sèvres after the death of Macquer, and Lavoisier, of whose qualifications no detail is necessary here, with Tillet, who had studied diseases of wheat, and Poissonnier, one of the King's physicians, who had experimented on the removal of salt from sea water. It was a well-knit committee; each member had been chosen for the expert knowledge that he could bring to its counsels. In 1786, other useful members were added.

The Committee appointed Lavoisier secretary—how often his biographers have to write these words!—and he proved, as elsewhere, its guiding genius. Immediately on the institution of the Committee, he obtained agreement for a number of proposals. He suggested that, as questions referred to it would not always need long discussion, members should give their opinions verbally. But when more important matters were raised that needed longer consideration, a member of the Committee should be charged with the duty of preparing a verbal or a written report for the next meeting; and when the Committee felt unable to deal with a question themselves, the Academy of Sciences should be consulted. He further proposed that a field for agricultural experiments near Paris should be obtained with all promptness, and he offered, if large-scale repetitions of these experiments were found necessary, to carry them out on his own land at Le Bourget, which he put completely at the Committee's disposal. He hoped that, presently, a journal for agriculture might be published. As one of the most active members of the Committee, it fell to his lot to examine the cloth bleached by chlorine according to the new process devised by Berthollet, to collect details on the cultivation of potatoes and maize, to draw up instructions on the folding of sheep, and to present reports on the use of turf ashes as manure and on the methods recommended by Tillet for preventing smut in wheat.

Lavoisier's economic views are well set out in a memoir that he composed when a question was put to the Committee by the Intendant of Languedoc. He reveals himself as entirely opposed to the excessive control of trade. The Intendant ascribed the scarcity of cattle in his province to a decree of 1763 authorizing export; and he asked the Committee to obtain the re-enactment of the prohibitive decree of 1740. Lavoisier argued that these prohibitions were always useless. When there was abundance, they prevented the disposal of the surplus, and in time of scarcity, conditions were against export, since prices were higher in the very area of scarcity. The Intendant asked also for prohibition of the slaughter of lambs and calves; but Lavoisier replied that the State had no means of insuring obedience to such a regulation, and did not know how to compel the rearing of animals when there was no fodder for them to eat. "Increase in price," he said, "decreases consumption, accelerates the return of plenty, and acts as a premium to attract supplies even from foreign sources, and such a premium has the advantage that it costs the State nothing."

He criticized control in problems of food supply and declared himself opposed to the disastrous policy of the Government which strove to reconcile in Paris two incompatibles in times of scarcity, namely, cheapness and plenty, "If, after making a first mistake in fixing too low a price for meat," he went on, "an attempt is then made to restore the situation by compulsory means or by resorting to that almost invariably ruinous expedient of financial companies, who never fail to offer themselves to the Government in times of difficulty and who have the art of disguising their own interests under the appearance of the public good, it will be felt that we cannot too soon revert to the principles that it has been sought to establish, by destroying every trace of the Government-authorized company, by raising the price of meat in Paris, even a little above its proper level, and perhaps by giving premiums for the importation of foreign beasts. This is precisely the plan adopted during recent years for wood. As long as the price was kept below its proper figure, no wood was brought to Paris; as soon as this was adjusted, there was abundance." As for selling meat above the price fixed by the Government, this, although it might have its advantages, could not be done in present circumstances. The butchers of Paris were a small corporation with the exclusive right to sell meat and, without a fixed price, the public would be delivered to all the abuses of a monopoly.

Finally, he concluded that the best way of remedying scarcity was for the State to interfere as little as possible. Whenever it tampered with the control of prices of consumable goods, it prolonged scarcity instead of remedying it.

The Committee resolved to institute in Paris a school of weaving to train pupils, who would go into the provinces to develop similar schools. France produced considerable quantities of flax and hemp, but exported them to foreign manufacturers; the people of France then bought the products into which their own native raw materials had been made. Lavoisier proposed to set up mills to use these products of the soil, to free France from the tribute thus paid to the foreigner, and to find work for the country folk during the season when they were not occupied with work in the fields.

As no help could be expected from the debt-ridden State, members of the Committee formed, as a charitable institution, a society with a capital of 9,000 livres, divided into shares of 300 livres among twenty-one members. The workshop was set up in the new boulevard, near the Rue du Montparnasse, and M. de Vergennes and the curé of St.-Sulpice devoted much of their time and attention to it. Lavoisier watched its progress: "I have followed all the details of this manufacture," he wrote, "there is not one of the pieces that I have not seen on the loom in the factory at Montparnasse." Moreover, he was the first to apply on a large scale in this manufacture the process of bleaching by chlorine.

The Committee appointed Lavoisier to draw up and edit the instructions on agriculture that the King had commanded to be prepared and sent to the Provincial Assemblies. The *Instruction* (1787) asserted that agriculture was the basis of national prosperity and pointed out that the restricted number of cattle was the principal obstacle to progress in a cultivation that disposed of an insufficient quantity of manure. It recommended that the Provincial Assemblies remedy this defect by furthering the practice of making meadows, supplying cultivators with seed free or on loan, giving cattle to farmers who had established a certain acreage of meadow, encouraging the production of winter fodder, beets, turnips, and potatoes, spreading a knowledge of the folding of sheep, and indicating the kinds of stock suitable for improving the breed and the quality of the wool.

To encourage the Provincial Assemblies to adopt these plans for progress, the *Instruction* emphasized the general well-being that would result from the increase of stock by the greater yield of food products, milk, butter, and cheese, and also meat, which would then come within the reach of the better-off country folk, while the hides and fleeces would provide raw materials for tanning and weaving.

The Assemblies were instructed to find out whether it was advisable to use horses instead of oxen for ploughing in some parts of their provinces, and to plough and harrow and then roll after sowing, where the land was suitable, instead of making deep furrows and thereby wasting land. These changes would save time and money both in seed time and harvest, in the latter by making possible, as in Flanders, the use of the scythe on the leveled field in place of the sickle on a deeply furrowed one. This practice would lead to the recovery of a greater proportion of the straw and also enable the farmer to take quicker advantage of good weather for reaping. They were also to give their attention to the cultivation of oats, the advantages or disadvantages of harvesting this crop in sheaves, diseases of wheat and their prevention by the methods so far available, improved processes of grinding to replace the inefficient and wasteful methods in use at that time, the draining of marshes, the maintenance of the navigability of the rivers, the cultivation of hemp and flax, and the manufacture of linen. The vast plan drawn up by Lavoisier for the Provincial Assemblies covered, in fact, the whole range of subjects and problems with which French agriculture was concerned in 1787.

Rural economics

OUTSIDE THE PURELY TECHNICAL ASPECTS OF THIS NATIONAL problem, agricultural progress was faced with many obstacles arising from economic conditions. Lavoisier dealt with these in a similarly exhaustive manner in a memoir at the meeting of July 31, 1787, presided over by the new Controller-General of Finance, Laurent de Villedeuil. When Calonne was dismissed from his Ministry on April 8, 1787, the Committee was doubtful of its continuation. Lavoisier, however, made representations to the new Minister, presented his colleagues, obtained sanction for keeping the Committee in being, and even persuaded the Minister to act as its President.

At this important meeting, with the new Minister presiding, Lavoisier began by pointing out that agriculture—which was first among all French industries, the gross value of its products amounting annually to nearly 2 milliards 500 million livres, and which contributed to the revenue, to the feeding and clothing of the people, and to the export trade—was in a state of inferiority compared with that of England. Too much attention had been paid to trade, and agriculture had made no progress during the century; an acre of English land yielded nearly twice as much as an acre of France. He said:

> But why is agriculture less advanced in France than in England? Frenchmen are neither less industrious nor less skillful than Englishmen; they succeed, as do the English, in all they take in hand; they nearly always equal and sometimes surpass the best that is done in

> England. Let us be bold enough to say that agriculture is a profession practised by the poorest class of the people and that, until the reign of Louis XVI, the people counted for nothing in France; it was only the power, the authority, and the wealth, of the State that were considered; the happiness of the people, the liberty and well-being of the individual, were words that never fell upon the ears of our former rulers, who were not aware that the real object of government must be to increase the sum total of enjoyment, happiness, and welfare, of all its subjects. If trade has gained the ear of authority and received more attention, it is because it is practised by a class of citizens with more education, who know how to write and to speak, town-dwellers, who form guilds and whose voice is more easily heard. The unfortunate husbandman groans in his cottage, unrepresented and undefended, his interests cared for by none of the great departments of the national administration.

Lavoisier enumerated the economic conditions that hindered the development of agriculture: (1) the *taille,* which arrested every kind of improvement because it brought down on the improver the burden of increased taxation and thus, in effect, acted as a premium for the discouragement of agriculture; (2) the humiliating *corvée,* which reduced the King's subjects to serfs, took men from the fields, often when they were most needed there, and suspended the very work on which the whole national wealth ultimately depended; (3) the *champarts,* a tax in kind levied by first choice of a part of the products of cultivation, together with lay and clerical tithes, which, in some parts of France, took more than a half and sometimes the whole of the net profit of cultivation; (4) the vicious nature of most of the taxes on consumable commodities; (5) the domiciliary visits for taxes on wines and other drinks, salt, and tobacco, visits that trespassed on the home and often led to inhuman, even indecent, searches, and brought the authority of the most humane of rulers into disrepute; (6) the *banalité* of the mills, by which people of the countryside were compelled to take their corn to a particular mill and were thus victims of the greed and monopoly of the millers, as well as the inefficiency of their mills which forced more than half the nation to consume food of poor quality and lost at least a sixth part of the flour owing to bad grinding; (7) the ancient right of *parcours,* by which, after harvest, fields were held in common for pasture and fencing was forbidden, and which therefore prevented enclosure of land and clearing of brushwood, led to loss of a second crop or lattermath and of part of the manure, took away from the cultivators any desire to make improvements, helped to communicate, spread, and propagate, epidemic animal diseases, and, lastly, disordered the land by the trampling of the cattle; (8) the assumption by riparian owners of the right to dam the streams in order to work their mills, with consequent flooding of good pastures and their conversion into marshes and the loss of a vast quantity of agricultural produce; (9) the prohibitions on export almost always decreed by the Government, which restricted production and in some measure prevented the growing of more corn than the kingdom could consume.

Asked Lavoisier:

> Is it credible that a realm so fertile, so essentially agricultural as France, which should be exporting all kinds of products, is short of hemp, flax, oil, wool, and cattle, which are imported in considerable quantities from without her borders, and that France is at the mercy of the foreigner for much of the agricultural produce for which her soil is most fitted? This condi-

> tion of decline and neglect in what is in France the first and most useful of all the arts, which employs most of her people, and which is capable of contributing most to the wealth and the power of the nation and especially to the well-being of her people, is due to this—that no one has applied himself to the problem, that the linking of agriculture with other departments of State has never been thought of, and that agriculture has remained a thing apart, without encouragement, without help, without funds.

Stressing again the need for the removal of the obstacles to agricultural progress that he had enumerated, Lavoisier added: "Instruction and encouragement are doubtless necessary; but neither instruction nor encouragement will succeed without the removal of the obstacles that bar the way to every improvement, every advance, every useful change."

He gave an account of the work done by the Committee during the previous two years, and its plans for the future. There had been correspondence with the Intendants of the various provinces and with agricultural societies, and especially with the curés of the country parishes. Rural associations with good promise of future success had been established for instructing the farmers and encouraging agriculture. A comparative table of all the varying measures of land, grain, and liquids, then in use had been started; instruction in the making of pastures, the cultivation of flax, turnips, and beets, the folding of sheep, and the liming of wheat, had been published. Work had been done on the improvement of stock and of wool, the setting up of spinning mills in the country for work during the winter months, the continuation of the rural and mineralogical atlas of France begun by Guettard, and the assembling of a collection of all agricultural implements, together with all the agricultural machines used in England and elsewhere. Taxes and legislation obstructing agricultural progress had been analyzed and the relevant reports had been prepared.

But in all this work they were hindered by lack of funds, especially for the free distribution of seed and of rams from abroad. And more than this was needed; for instruction was valueless without encouragement, precept without example. The establishment of experimental and model farms, both near Paris and also in the provinces, the work on measures, on the geological atlas, and on the collection of implements, had been held up for lack of money. In fact, two years had been lost because, after the previous Minister had enlarged his first views on the work of the Committee, the original grant-in-aid was in the meantime suspended without more ample funds being set aside for the larger purpose.

Now, however, said Lavoisier, was the moment for action. The drought of 1785 and the scarcity of fodder that had followed upon it had brought home to the farmers the need to make every provision against the consequences of another such scourge; and now was the time for distributing seed and establishing new crops. "The expenditure now necessary," he added, "to set up a Department of Agriculture is not incompatible with the spirit of economy that animates the present administration. It is on unproductive expenditure, on expenditure that yields and can yield no profit, that economies should be made. Expenditure for the improvement of agriculture is of another kind; it is expenditure for production; it is seed that may yield more than a hundredfold. It was shown, in the memoirs submitted to M. de Calonne, that the value of the gross products of agriculture amounts to 2 milliards 500 million livres. An improvement of only one-tenth would increase the national wealth by 250,000 million livres. Assuredly the King can make no better investment for the nation or for himself; for national wealth cannot be increased without at the same time increasing the revenue."

In conclusion, Lavoisier proposed the institution of a State Department for Agriculture in three sections: one for agriculture properly so called, with members from the Academy of Sciences and the Society of Agriculture; a second for trade; and a third for matters relating to the Tax-Farm. Members of the three sections would constitute a General Committee for agriculture, trade, and finance, under the Controller-General of Finance.

The Committee that heard these magnificent plans put forward ceased to function at the end of that year, 1787. It had done much to achieve its purposes, but it was helpless against the indifference of a centralized authority. In 1788, however, a decree, reorganizing the Royal Society of Agriculture, appointed a committee of eight, of whom Lavoisier was one and Dupont de Nemours another, to study questions of agriculture and rural economy on which the Government thought it proper to seek their advice. No trace of its existence has survived except the report, in Lavoisier's handwriting, of a meeting held in August of that fateful year, 1789.

XX

THE PROVINCIAL ASSEMBLY OF ORLÉANAIS

LOCAL GOVERNMENT AND ADMINISTRATION IN THE PROVINCES of France had long been entrusted to Intendants, or Provincial Governors, who, in general, knew little of and cared less for the needs or the sufferings of the people. Under the authority of a central power that had conferred on them an almost limitless jurisdiction they ruled as despots, free to commit every abuse and every injustice. "There is one cry, and one only, throughout the eighteenth century," wrote Léonce de Lavergne; "it is against the rapacious government of the Intendants; passive agents of a fiscal tyranny, that habitual scourge of absolute governments, they exhausted of men and of money the unfortunate provinces that had been delivered into their hands."

Necker became influenced by the ideas of Victor de Riquetti, Marquis de Mirabeau (1715-89) and father of the Revolutionary Comte de Mirabeau, as expounded in his *Ami des Hommes* or *Friend of Men,* and later by those of the physiocrat, Turgot. During his first tenure of office as Minister he decided to restrict the powers of the Intendants by handing over to local assemblies many of their most important administrative responsibilities: the partition among inhabitants of the provinces of the taxes due to the central government, the levying of taxes, the construction and maintenance of highways, the encouragement of trade and industry, and the exploitation of new markets for their produce and manufactures.

The first Provincial Assembly to be instituted was that of Berry in 1778. In it representatives of the three orders, Clergy, Nobles, and *Tiers État,* or Third Estate, sat together for their deliberations as one body, the number of representatives of the Third Estate being equal to the total number of representatives of Clergy and Nobles. Every representative's vote was therefore equal and there was no voting by orders in separate "houses," by which the Nobles or the Clergy, voting as an order, might reject a resolution passed by the Third Estate. Voting was by heads in one common council, not by orders in different chambers; this method was to become important, as we shall see, when the States-General met in 1789 for the first time since 1614. Two other Assemblies were established in 1779 and 1780, but Necker was dismissed in 1781 and the trend towards decentralization of government was halted. Not until 1787 did the Minister (Calonne), inspired by Dupont de Nemours, obtain approval for the institution of Provincial Assemblies throughout France.

The first session of the Provincial Assembly of Orléanais was arranged for September 6, 1787. Its constitution was fixed at twenty-five members chosen by the King—six clergy, six nobles, twelve commoners, and a President (the Duke of Luxembourg); these twenty-five members were to elect a further twenty-five in the same proportion for the three orders. Between sessions of the Assembly its decisions would be carried out by a permanent Intermediary Commission.

Lavoisier, through his purchase of Fréchines, was a landowner in Orléanais but, although a noble and ranked as an esquire through his father's purchase of an office that carried the privilege of hereditary nobility, he was elected as a representative of the Third Estate for the little town of Romorantin. This was not unusual; nobles and clergy were often elected for the Third Estate. With Mme. Lavoisier, he left Paris for Orléans on the evening of September 4 for the opening of the Assembly on September 6. The first meeting, as described by Grimaux, was a ceremonious affair, at which the Intendant of Orléanais, as the King's Commissioner, inaugurated the Assembly "in His Majesty's name." He was received by four representatives of the Assembly—the Abbé de la Geard from the clergy, de Salaberry from the nobles, and Lavoisier and de Bonvalet, Mayor of Orléans, from the Third Estate. After being conducted to his place, he announced the King's wishes and indicated the items to which the Assembly were to direct their attention during this first session, namely, the formulation of rules of procedure and the nomination of the further twenty-five members to be elected. These tasks were the only duties prescribed for the session. The President made a brief reply: and the King's Commissioner was escorted from the Assembly with the same honors and by the same representatives as had greeted him on his arrival. The Assembly then adjourned until September 10.

When the Assembly sat again, four days later, Lavoisier presented a report on the preparatory work to be done by the Intermediary Commission in readiness for the second session of the Assembly due to be opened on

November 17. Rules of procedure were drawn up and the twenty-five additional members were elected. Among them were the Abbé Sieyès, who during the Revolution would devise a multiplicity of unaccepted constitutions for France and vote for the death of Louis XVI, and Lieutenant-General the Comte de Rochambeau, who had commanded the French troops sent to America. On September 14 the Assembly completed its first session, which was closed by the Intendant with a repetition of the previous ceremonies.

Lavoisier hurried back to Paris and left there on September 22 with two colleagues from the Academy of Sciences, Monge and Fourcroy, for the iron works at Le Creuzot, on which they were to make a report. He returned on October 3, for work with the commission reporting on the hospitals and with another commission of the Academy concerned with the judging of entries for the prize to be awarded for dyeing. These duties done, he set out again for Orléans for the second session of the Provincial Assembly. There he displayed again his remarkable powers to influence others, his astonishing capacity for work, and the dominating superiority of his talents; he became the guide of its deliberations. "It was Lavoisier who did everything," wrote Léonce de Lavergne," who inspired everything, who was everywhere."

The second session was opened, again with ceremonies, says Grimaux, on Saturday, November 17. The Intendant indicated the matters to be dealt with—a more equitable arrangement of taxes, reform of the *corvée,* the development of new manufactures, and so on. The President made a formal reply; and then the Intendant delivered a lengthy royal instruction, regulating the precedence of the different subjects to be dealt with and the conduct and duties of the Assembly, and ordered the nomination of the

Intermediary Commission, as well as two procurators to carry out the duty of presenting to him a concise report of the proceedings "in order that His Majesty's Commissioner may be assured that no subject is discussed other than those with which it is the duty of the Assembly to concern themselves." Next day, Sunday, the Assembly attended the celebration of Mass, which, in accordance with the King's commands, must precede their labors. The members gathered at the bishop's palace and marched in procession to the cathedral—Third Estate first, as being the lowest in the social grade, headed by a military escort and the city band. The clergy and the nobles followed, each order abreast of the other, as befitted social equals, the clergy on the right and the nobles on the left; then came the President, the Duke of Luxembourg, completing the procession. Mass was celebrated by one of the clerical members of the Assembly; and a sermon on the love of one's native land was delivered by another.

Vingtièmes

ON MONDAY, THE ASSEMBLY SET TO WORK ON ITS TASKS, which were divided among four committees, the first concerned with social conditions and agriculture, the second with bridges and highways, the third with accounts, and the fourth with taxes; a special committee was appointed to study problems relating to the tax of *vingtièmes,* or "twentieths," levied on property. Lavoisier served in the first section, which consisted of ten members, five from the Third Estate and five from the nobles and the clergy, these including the Comte de Rochambeau and the Abbé Sieyès.

The first proposal to be discussed by the Assembly,

states Grimaux, was the commutation of the *vingtièmes* into a fixed sum to be paid annually to the Government. At once the clash between the privileged classes and the national interest became apparent. Lavoisier was not a member of the special committee concerned with discussing the *vingtièmes,* but by his influence on the representatives of the Third Estate, by the memoirs that he drafted and edited, and by similar activities, he played an important part in the study of this problem, on which the opinions of the Assembly were divided.

Vingtièmes had been originally imposed as a tax on property in 1710. At first the tax amounted to one-twentieth of the return on the property, then two-twentieths, and then four sols in the livre, which made it four-twentieths. The national revenue from this source in 1784 amounted to 76 million livres, and in 1787 in the province of Orléanais it amounted to 1,800,743 livres. The small landholder found it a burdensome tax; the nobles were either exempted from payment or paid at a reduced rate. It varied and it increased every year; over a period of sixteen years in Orléanais it had increased by 400,000 livres. The proposal now before the Provincial Assembly, by commuting the tax into a fixed annual payment, would eliminate all fear of an increase in the amount and, further, would enable the province to collect its own *vingtièmes,* apportioning the payment of tax between the different sections of the inhabitants on more equitable lines than had hitherto been followed by the fiscal agents of the central government. The fixed sum proposed by the Government was 2,500,000 livres. Although this amounted to much more than the current yield of *vingtièmes* from the province, there was some advantage to be gained by accepting the offer, since reapportionment would provide some relief for the poorer farmers. The proposal would also apply the tax to much property that had previously been exempted from it, including the estates of the King, and of Monsieur, the King's eldest brother, as well as lands of the nobles.

There was much animated and excited discussion among all the representatives. The response turned not so much either on the principle of commutation or on the sum named as on the liability of the nobles to be taxed for *vingtièmes*. The Third Estate wished to accept the offer, especially when examination of the returns showed that the Duke of Luxembourg, a great landowner in the province, paid no *vingtièmes;* that the Baron de Montboissier paid on an income of 12,000 livres, whereas his lands brought him 60,000 livres; and that de St.-Fargeau had obtained a reduction of 1,500 livres on a sum of 3,000 livres, while the tax was rigorously exacted from the poorest of the King's subjects. Almost all the nobles and a few members of the Third Estate, were against acceptance, while the clergy, whose interests were not at stake, favored the proposal. The special committee on *vingtièmes* were evenly divided; the Duke of Luxembourg and the Baron de Montboissier were opposed to the proposal, while it was favored by the Abbé Sieyès, the Bishop of Chartres, and the Comte de Rochambeau. The latter, unable to attend because of illness, wrote, according to notes of the work of the Assembly drawn up by Mme. Lavoisier and quoted by Grimaux, to advise his fellow-noblemen: "The nobility can gain the respect of the province only by supporting the interests of the province and by abandoning their privileges for the general good of the nation."

Finally, on November 30, the Assembly accepted the King's proposal for a fixed sum, but resolved to offer 2,300,000 instead of 2,500,000 livres, at the same time requesting that the necessary administrative work should be handed over to the Assembly, which would reduce the costs of collection. The Government adopted these proposals early in 1788, the new system to become operative on July 1.

Corvée

The *corvée,* the system of forced free labor on the construction and maintenance of roads and the compulsory provision of horses and vehicles for transport, had been abolished in 1776 by Turgot. Having to find money to pay for road labor, he raised it by a further tax on all those who paid *vingtièmes.* The nobles protested at their having to pay for work on the roads, which had hitherto fallen on the commonalty, arguing that it led to a confusion in the order of society, the nobleman being no longer distinguished from the commoner, the lord from the peasant. When Turgot fell after a brief period of office, his successor suspended the decree abolishing the *corvée.* Forced road labor was continued until June 7, 1787, and the costs were then raised by a tax from all who paid the *taille;* the nobles thus still escaped payment.

Lavoisier was stirred by this long continued injustice, and proposed in the committee on social conditions and agriculture that the Assembly should urge upon the Government the reintroduction of Turgot's decree. He presented a memoir on the subject, reminding his fellow-members of the importance of highways and roads in increasing the value of all lands that bordered upon them by reason of the easier outlets that they gave for all traffic in produce and commodities. Money spent on roads was a profitable investment for landowners; and a road was practically valueless and a waste of capital until it was completed to its last league. It was to the advantage of the landowners that roads should be made and that, once begun, they should be rapidly completed so that the expenditure on their construction might be a source of profit to the public as soon as possible. But, while rapid construction was in the interests of the landowners, the interests of all those paying *taille* (by a part of which tax the revenue for making the roads was raised) must be considered so that they were not taxed beyond their resources; and it was, said Lavoisier, his intention to suggest a means of reconciling these opposing interests.

The Comte de Rochambeau had already proposed that the tax replacing the *corvée* should be reduced from one-sixth to one-eighth, and it seemed that this must be done because the taxpayers were so heavily burdened that some relief was necessary. Meanwhile, the time necessary for construction of the roads had been estimated at seventeen years and, Lavoisier went on, the reduction now proposed in the tax would extend this to twenty-five years; and landowners would have to wait for this long period to elapse before they could expect an increase in the value of their property. But, argued Lavoisier, since those paying *taille* must clearly have some reduction in tax and since the landowners had a pressing interest in accelerating the construction of roads, why should the latter not offer to bear a part of the expense themselves and why should they refuse to agree to take over payment of the total amount involved in granting relief to those paying *taille?* He could not see the grounds on which landowners could complain about this arrangement, since they would find adequate compensation for their outlay in the increased value of their lands and in the rise in the prices of leases that would inevitably follow.

"These considerations," he continued, "have led me to

make some researches on the origin of the *corvée,* and the result has shown that there is neither in our history nor in our national law any ground for casting upon a single class of taxpayers a burden that, in justice alone, should be borne by all and perhaps ought preferably to bear more heavily on those to whom it is most profitable." He added:

> This Assembly is entirely composed of landowners, of whom by far the greater part do not pay *taille.* We must therefore be bold enough to take upon ourselves at this moment the defense of the cause of those paying *taille,* especially in a tribunal where the interests of almost all its members lie in the opposite direction. But I address myself to this task with confidence, being persuaded that the upper classes of the nation and all those who share their privileges will be the first to reject every exemption that they do not think to be justified and that, in this century of enlightenment and beneficence, they would not wish to put upon the hard-working and indigent section of the nation, upon the class that produces all its sustenance and all its wealth, a burden that, according to the ancient laws of our constitution, should be borne by all.

Lavoisier summarized his historical researches on the *corvée* and showed that, until the recent declaration of June 7, there had been no legally established exemption from taxation for the cost of the construction and maintenance of roads; that such an exemption existed neither under the Roman emperors in any part of their empire, and therefore in Gaul, nor under the Kings of France; and that it had now been made in fact but not in law, *de facto* but not *de jure,* without the authority of any legislation. Moreover, there had been no exemption either under the ordinances of Charlemagne or the letters-patent of Henry III; and Louis XIV had on July 18, 1670, on the advice of Colbert, directed "that the highways and side-roads shall be kept in constant repair and maintained at the charge and expense of the landowners of the parishes where there are bad highways, with stones, gravel, or faggots of brushwood, according to the Statutes."

It was then clearly not a question of making a new law, but of insuring obedience to an existing one, since the phrase ran "according to the Statutes." There was, therefore, no question of any exemption or of making the commonalty exclusively responsible for the cost of the work on the roads. But, in 1720, some of the Intendants of provinces, after the disasters of the War of the Spanish Succession and the famine of 1709, realizing the harm that would follow to the nation if the roads were neglected, and faced with the necessity of repairing them without the hope of any help from an exhausted Treasury, introduced for the first time the application of the *corvée* to work on the roads. There had, of course, been compulsory work on manorial roads and on military roads, but no law had ever established it for the King's highway. Besides, for the *corvée* on manorial roads, there had been grants of land in return; and as for the military *corvée,* it was but one more act of hostility in enemy territory, while, inside the frontier, the sovereign law of the safety of the State justified this violation of personal liberty.

Continued Lavoisier:

> But that in time of peace, in this very century, without law or warrant, it should be permitted to take the laborers from their work in the fields, that there should be exacted from them what one would have scarcely dared to demand from serfs, that this innovation should have aroused no complaints or protests

on the part of the courts, that a half-century later it should be argued that those who pay *taille* are naturally subject to *corvée,* that this is the essence of the constitution of France—this is what posterity will one day find it difficult to believe. . . . The practice of making splendid public works, without directly demanding anything from either the great or the wealthy, whose voice alone had counted up to that time in France, of making a name for oneself, of acquiring friends and protectors, while forcing the poor to level the roads traversed by the well-to-do, soon became general. It seemed easy to create roads by statutes. It was self-deception to think that they cost the State nothing. At the time when the method of ordering *corvées* for the making of roads was adopted, it was expedient to exercise self-restraint by placing this public charge solely on the people. None ventured, and none could dare, to send the nobleman or the priest on *corvée,* because their rank protected them from servile labour. It was then that our constitution was changed on this point and that the two upper orders of citizens gained *de facto* exemption from contributing to public works, and that the Statutes of our Kings, constantly renewed before and since Charlemagne, were set at naught. But neither the ordinances of the Intendants nor even the instructions that they have been able to obtain from the Council with regard to work on the roads, when *corvées* were generally employed for this work, have ever been laws of the realm; they have received no sanction; they cannot therefore supersede a national law, more ancient even than the monarchy. . . . The edict of 1776 is the first law of the realm in which the *corvée* for the roads is mentioned and it is named but to abolish it. . . . Do not doubt that the honorable step that I suggest that you should take to secure the re-establishment of the ancient laws will receive the approval of the Sovereign. . . .

Lavoisier then proposed that the Assembly should petition the King for a law to distribute the burden between all classes, pointing out that such a practice was already operative in some provinces. He appealed to the privileged orders in these words: "We are therefore of the opinion that the Assembly should with all promptness present a petition to the King for the restoration of the ancient national law, and that the entreaties of the nobles and the clergy in this matter should be, if such were possible, even more urgent than those of other citizens, since honor prescribes the more vehement rejection of an injustice that is profitable than one that is injurious."

Lavoisier had even drafted the necessary resolution. It read: "Resolved that His Majesty be very humbly petitioned that he be pleased to order, by means of a statute to be registered, that the payment of the taxes assigned for the work on the roads shall be shared by the landowners of all orders, in the same way as was done for the construction and repair of the presbyteries, and this, not as the result of a general deliberation, but as a resolution which was formulated mainly by the Third Estate and which all the members of the Clergy and Nobility can but individually commend."

For his proposals, which so far had been made only in committee but not before the Assembly, Lavoisier obtained support from some of the nobility, including the Duke of Luxembourg, the Comte de Rochambeau, and others. There was opposition, but by a majority it was resolved that the memoir should be read to the Assembly; and the representatives of the Third Estate warmly congratulated Lavoisier on his devotion to their interests.

Later, most of the nobles and some members of the Third Estate, on second thoughts, decided to make formal objection to the reading of the memoir when the time came, on the grounds that the subject had not been included in the agenda for the session. The committee, probably taking fright at their resolution that the memoir should be read to the Assembly, sought to escape responsibility, and adopted the view that there had been no formal resolution and that they had authorized the reading of the memoir only as a personal contribution of the author and not as their collective opinion. Faced with this opposition, Lavoisier, according to a manuscript note by Mme. Lavoisier, withdrew his memoir, despite the advice of friends who had at first hoped for a majority in the Assembly. Nobles and clergy had combined to oppose a measure that encroached upon their interests. The records of the Assembly make no mention of this important memoir, but the manuscript in Lavoisier's handwriting is preserved in the Library of Orléans.

Social security

THE MOST IMPORTANT DUTIES OF THE ASSEMBLY WERE IN the hands of this same committee and almost all the reports were the work of Lavoisier. He presented once again his views on agriculture in France and then dealt with the special problem of agriculture in Orléanais. He read memoirs on trade, on charitable institutions, and on tariff duties at the boundaries between provinces suggesting for the latter the commutation of these dues for a fixed annual sum.

He proposed a scheme for insurance against poverty and old age, opening his memoir thus:

Man, at the moment of his birth, is absolutely without power to satisfy his needs; he is in dependence on others and he can survive only by the continued care that they take of his existence. Gradually age brings about the development of his strength and there comes a time when that strength not only suffices for his own needs, but is also sufficient for him to use it in giving to others the help that he formerly received. This period is the best in his life, but it is limited in duration; at the approach of old age, his strength gradually decreases, as it increased in youth; the aged fall back into the same state of dependence in which they were in their infancy, and with the greater regret for the pleasures they have lost and the sad recollection of what has been and will no longer be.

Happy is he, at this time, who is the father of a numerous posterity! Happy is he who, surrounded by a grateful and tender family, receives from them in his last years the help that he lavished on them in the vigor of his younger days, and who is led to the end of his course through a quiet and tranquil old age! But such happiness in the last years of life is not given to all men: some grow old without having had the good fortune to see themselves live again in their children; others have had the misfortune to lose them at the very time when their help was becoming more necessary. They have seen the loss, and in a moment, of the fruits of many years of toil and labor; some, still more unfortunate, have only unnatural children who abandon them and repay them only with ingratitude. But, without bringing accusations against mankind, without stopping at such instances which happily are rare, do we not see every day that an unfortunate laborer with a numerous family gains, even by exhausting his strength, scarcely enough bread for his wife and children? And by what right could we demand that he should prefer those who have brought him into the world, but who are now fallen into senility, to the companion that he has chosen, to the mother who has borne and nursed his children, to the very children to whom he has given life? And who shall venture to decide between the choice of giving or of refusing to such loved ones the sustenance that is necessary to all, but which is already insufficient for the family?

We are not unaware that hearts that are unmoved at the sight of suffering humanity look upon an old man, condemned by feebleness to idleness, as a burden to society which it is to the common interest to be rid of; and that they will be little touched by the care that we are taking to obtain support for this class of unfortunates. It is not to those hard hearts that we are addressing ourselves. The burning zeal that animates you for the well-being of humanity, the spirit of patriotism with which you are penetrated, reply in advance that there are none such among you.

Accordingly, Lavoisier proposed a scheme of insurance against poverty and old age, and a savings bank with annuities based on the mortality tables published in 1785 by de la Roque. The actuarial details of the scheme were carefully computed and are anticipations of the later ideas of our own time on social security. "We propose, therefore," he said, "to institute at Orléans, under the title of *Caisse d'épargne du peuple* [People's Savings Bank], an establishment to receive the sums contributed by persons of every age and condition who wish to insure for themselves, their widows, or their children, an annuity determined according to the tables that have been drawn up for this purpose. The whole province will be the guarantor of the commitments undertaken by the bank and of all the actions that are carried out in conformity with the regulations that will be prescribed for it."

This is the primitive form of our modern systems of contributory State insurance. It is interesting to see what followed. A committee was appointed to examine the matter; but before their deliberations were complete, a finance company had obtained a privilege for fifteen years for setting up such a scheme, basing their prospectus on de la Roque's tables. De la Roque's comment was that it had not occurred to him that what was essentially a work of benevolence could become a matter of financial speculation and that it would be preferable to pass the profits to the hospitals. Lavoisier felt that the plans of the committee had been anticipated and that, if these facilities were available to the people and if the company were founded on an equitable basis, it mattered little by whom the scheme was put into operation.

Other recommendations and reports

LAVOISIER ALSO PRESENTED PROPOSALS FOR THE ESTABLISHMENT, under the authority and control of the Provincial Assembly, of charity workshops to give employment to those not fit for full-time labor and to the genuinely poor, and of other workshops for vagrants in order to reduce the numbers of those who made begging their trade. These suggestions were received with approval by the committee.

During these discussions the Orléans Academy and the Agricultural Society invited the Provincial Assembly to prescribe subjects for the annual prizes for their essay competitions. For the former, Lavoisier's suggestion was the institution of the Savings Bank, and for the latter, the best means of re-apportioning the payment of taxes, both personal taxes and property taxes, such as the *vingtièmes* and the *taille* and other accessory taxes, without theoretical treatment, but with special reference to the province of Orléanais. These suggestions also were adopted.

Further, Lavoisier drew the attention of the Assembly to the scandal of the high mortality among foundlings put out to nurse in Paris, only three in a hundred of whom reached the age of six years. He wrote to Vic d'Azyr, Secretary of the Royal Society of Medicine, for advice on the education of children without nurses and he obtained from his colleague, de Salaberry, the financial help for a new practical study of this problem.

When the Duke of Luxembourg, President of the Assembly, proposed the preparation of a map of Orléanais, Lavoisier added a further proposal for the inclusion of geological details in the map. He offered to do the work himself, as he had Guettard's information at his disposal, together with other data that he had acquired later, and he could obtain still further details from scientists and engineers in the province. The Assembly agreed also to his proposal to make a collection of the mineral products of Orléanais, including even the commonest.

Other reports that he presented dealt with the improvement of the quality of wool, the navigation of the Loire, the institution of a discount bank at Orléans to increase the available credit for trade (France being very short of capital), and the reduction of the expenditure on the collection of the taxes by the suppression of costly intermediaries.

Before adjourning, the Assembly nominated Lavoisier as one of the six members to serve on the Intermediary Commission. This Commission continued to meet until the end of 1790, but the Provincial Assembly never held another session and none of the reforms that it advocated was ever achieved. All these generous and competent activities, these able and devoted labors, were a complete loss. Events began to move too quickly to a climax. Evolution gave way to Revolution. The publication of a Royal Edict in August, 1788, for the assembling of the States-General in the following year dwarfed all interest in these local activities. The work of the Provincial Assemblies was overshadowed by the importance and urgency of the national summons.

XXI

RETROSPECT—1787

IN 1787 LAVOISIER WAS IN HIS FORTY-FOURTH YEAR WITH only seven years of life before him. He had distinguished himself as chemist, geologist, cartographer, economist, agriculturist, social reformer, Academican, Farmer-General of Taxes, and Commissioner for Gunpowder, and presently he was to formulate a national system of education. He had won a secure place in science and in the public affairs of France; and rarely can a human life have been so full. But we inevitably ask ourselves, "What manner of man was this?" What other interests had solaced him and refreshed his spirit through these years of tireless energy and remorseless activity? What friendships had sustained him?

No complete answer can be given to these questions, because Lavoisier's vast correspondence, still largely extant, is only now being edited and prepared for publication by a committee appointed by the Academy of Sciences, of which he was one of the greatest ornaments in its long history of nearly three hundred years. However, the preliminary handlist compiled by Professor René Fric shows that, apart from the letters that passed between Lavoisier and his father and his aunt during the geological tour with Guettard in 1767, the bulk of his correspondence relates to his scientific and official activities and gives no information either about his friendships or about his distractions other than what was already collected by his painstaking biographer Grimaux, more than sixty years ago. As Lavoisier was not given to destroying papers, but rather to the preservation of even hurriedly written notes and drafts, it is probable that little is missing from the surviving documents, and that we are not likely to have much more information than the fragments that have long been available.

The earliest friend of Lavoisier of whom we have any record is de Troncq, whom we have mentioned earlier, the young man who sent "my dear mathematician" a letter of advice, accompanied by a basin of gruel. All that survives of this early friendship is a couple of letters and we know no more of the waggish de Troncq than this amusing episode. We find no trace of any other friends for some time.

In his home, the young Lavoisier seems to have been content. The devotion of his aunt and the extraordinarily close relationship between his father and himself may have made him a man of few intimates or even of none. His father's sudden and premature death in 1775, at the age of sixty, had deeply grieved him; to the Comte de Tressan he wrote that the mutual confidence between his father and himself had been one of the happiest things in his life.

The early years of his manhood had been busy years. In 1768, at the age of twenty-five, he had been elected to the Academy and had entered the Tax-Farm; and then came his marriage in 1771, at the age of twenty-eight. There may have been no time for social distraction, even for friendship, but, as we have already seen, Lavoisier had from his first acquaintance with science enjoyed the company of many older men, such as Guettard and others. It is known that he had been on the closest terms of friendship with Trudaine de Montigny, who was about ten years his senior and the son of Daniel Charles Trudaine, Counsellor of State and in charge of the Government Departments concerned with commerce and roads and bridges. When the father died in 1769, the young Trudaine de Montigny had succeeded to his father's office, but declined to accept the emoluments, a rare event in those days. He was much interested in chemistry and constructed a laboratory at his country house, Montigny, in Brie. Writing to Lavoisier on July 14, 1772, he said: "I am working hard to become your colleague in chemistry. I now have a laboratory and have not been outside it since it was finished three days ago." Lavoisier had visited his friend in 1774 and worked in his laboratory; but de Montigny died suddenly in 1777 in his forty-fourth year.

One other friend, a distant relative, was Augez de Villers, of whom, however, almost nothing is known; but it was to de Villers that Lavoisier wrote his last letter on the eve of his trial before the Revolutionary Tribunal with his fellow members of the Tax-Farm. This letter will be quoted later.

Lavoisier seems to have been a man of many friends but, so far as we know, with no close friend. He had long been well known and frequently in conference with the Ministers of State, Turget, Necker, and d'Ormesson. He had had close contacts with scientists from other countries, including Benjamin Franklin, who was in Paris as diplomatic representative of the United States of America from 1778 to 1786. After his removal to the Arsenal in 1775, his laboratory had become a rendezvous for the whole world of science and there were few eminent scientists who had not visited him there, witnessed and even taken part in his experiments, and enjoyed the gracious hospitality of Mme. Lavoisier.

Academicians and statesmen had been among his guests. To these, as well as to foreign visitors, Lavoisier took delight in showing the instruments and apparatus that had been made in France, since it was one of his constant aims to encourage this highly skilled work, for which France had previously been dependent on England. He had often helped in this development by advances from his own resources when the construction of an exact scientific instrument was beyond the means of the craftsman. His foreign visitors had included Joseph Priestley, James Watt, Charles Blagden, Smithson Tennant, and Arthur Young from England; Jacquin and Ingen-Housz from Vienna; Fontana and Landriani from Italy. Macquer, de Morveau, d'Arcet, Bucquet, Cadet, Berthollet, Vandermonde, Cousin, Laplace, Lagrange, Monge, and many others from the Academy of Sciences, had frequently come to his laboratory, together with the Duke de la Rochefoucauld, who had studied the formation of saltpeter on his estates at Roche-Guyon, the Duke de Chaulnes who had carried out researches on fixed air, and the Duke d'Ayen, President of the Academy of Sciences.

Young men who were interested in science had been welcomed to the Arsenal laboratory and allowed to help with the researches and introduced to the scientists who came there to carry out their own experiments. Among these younger men were Gengembre, who discovered the spontaneously inflammable gas, phosphoretted hydrogen;

Hassenfratz and Adet, who devised a new scheme of chemical characters or symbols; and the younger son of Dupont de Nemours, Eleuthère Irénée, who was later to be the founder of the great chemical industry of America. All experiments were repeated before Lavoisier's fellow chemists from the Academy and foreign scientists, if any were visiting Paris, before being reported to the Academy; criticism was invited and all theories were subjected to close discussion. For new and especially important experiments, invitations were despatched, and the company which was assembled included nobles and commoners. Here also for the convenience of all concerned Lavoisier had held meetings of the Academy committees to which he was appointed.

Of interests other than science and the public affairs of France, Lavoisier seems to have had few. But in 1778, after he had inherited a house at Le Bourget from his mother's family, he apparently sought closer contact with the country, for he had then bought the château and estate of Fréchines, where he began his agricultural experiments. He was fond of music and the papers that he left include a treatise on harmony. He had a box at the Opera; in a pamphlet of 1789 attacking the Farmers-General, this was the only item for which he was reproached. His life evidently was passed entirely in work. That may be the reason why the various chroniclers of eighteenth-century scandal, so formidable in their number and their productions and so varied in their charges—especially with regard to the Farmers-General and other financiers—have not one word to say about Lavoisier or, indeed, about Paulze, his father-in-law. He was interested also in art and a catalogue of an exhibition of painting and sculpture held in 1785, found among his papers, bears in his handwriting in the margins and on the blank pages a number of critical notes on the exhibits. Of the painter of a picture of the restoration of Alcestis to her husband by Hercules he wrote: "Either this man is not married or I pity him; he has never felt the joy of reunion with a loved one after separation."

In 1787 Lavoisier had completed twenty years of high endeavor and exacting toil, years marked by that astonishingly unflagging energy that characterized his whole life. The art of experiment is ever difficult; for, while experiment may be regarded as a means of asking questions of Nature, the questions have first to be formulated. As Robert Hooke, himself a great experimenter, wrote in 1665 in his *Micrographia,* "the footsteps of Nature are to be trac'd, not only in her ordinary course, but when she seems to be put to her shifts, to make many doublings and turnings, and to use some kind of art in endeavouring to avoid our discovery." Kindly Mother Nature has often seemed a jade to the scientific experimenter. But, after genius has posed the right questions in the right way, the answer is given; and a discovery that once seemed beset with difficulties and uncertainties then becomes a mere item of common knowledge and almost obvious in its apparent simplicity.

Methodical habits go far to explain the number, although not the nature, of Lavoisier's many achievements during these twenty years. He customarily allotted six hours a day to work in the laboratory, from six to nine in the morning and from seven to ten in the evening, leisure hours outside his "working day," which was otherwise occupied with heavy duties in his various capacities. After 1775, he had lived at the Arsenal and from Mme. Lavoisier we have a glimpse of his life there. One day a week was wholly given up to experiments. "It was for him," she wrote, "a day of happiness; some friends who shared his views and some young men proud to be admitted to the honor of collaborating in his experiments assembled in the morning in the laboratory. There they lunched, there they debated; and it was there that the theory that immortalized its author was created. It was there that you should have seen and heard this man with his precise mind, his clear intelligence, his high genius, the loftiness of his philosophical principles illuminating his conversation."

We may picture him also returning from a meeting of one or other of the many commissions or committees on which he served, generally as secretary. He sat down at his desk to draft his report and he redrafted it; he corrected it and re-corrected it until he was satisfied that it was an exact and accurate and clearly composed record. He did the same when he was preparing his memoirs. Then a fair copy was made by a secretary. But even then, Lavoisier was sometimes not entirely satisfied and his papers show that he would occasionally insert or delete or modify a word or phrase so that every detail was clear and certain. The style of his memoirs is clear, nervous, and graceful, and bears all the marks of one who knew how to use and handle words. He left so many papers that it is scarcely credible that he found time to compose so much. Kindly and courteous, he could write incisively and speak sharply.

Lavoisier's family had given several priests to the Church, including his great-uncle, Father Laurent Waroquier, one of his godfathers, for whom he had been named. In that age of atheism, Lavoisier kept his Catholic faith. When Edward King sent a copy of one of his books, Lavoisier wrote in acknowledging the gift: "You wage a noble cause in defending the revelation and the authenticity of the Scriptures: and what is remarkable is that you now use for defence the very weapons that have many times been employed in attack."

His probity as an official was well known to his contemporaries. When some Breton deputies urged him to appoint a Breton as gunpowder commissioner at Rennes on the grounds that a native was preferable to a stranger, he fearlessly reminded them that he would have opposed such a demand even from the representative of a dreaded and arbitrary government, since an experienced and qualified citizen who was justly entitled to the post could be no stranger, while the real stranger would be one who, lacking the necessary experience and knowledge and relying on the support of recommendations only, might come to occupy the place that was rightly due to another.

His deep love of humanity had long been evident from the many social reforms that he had suggested and planned, and especially in his attitude to the freedom from taxation enjoyed by the privileged orders. "If it is allowable," he said, "to make exceptions in favor of any class, especially with regard to taxes, it can only be in favor of the poor."

The same humanity towards his fellow men was displayed on many different occasions. He had helped younger scientists from his ample purse. He had helped others, too; and in his papers receipts were found for such loans amounting to over 80,000 livres that had not been repaid. At Fréchines he was the devoted squire of his people, settling their disputes and helping the sick, not only with gifts, but also with words of encouragement. He opened the first school in the neighboring village of Ville-francoeur; and he sold the produce of his farm at less than market prices, so that those in need might buy and thus avoid the feeling that they were objects of charity. In the famine that followed the bad harvest of 1788, he loaned money to the towns of Blois and Romorantin to buy corn, and refused to take any interest on the loans. When he was in charge of the district of Clermontois as an official of the Tax-Farm, he freed the Jews, as we have seen, from the degrading tax of *pied fourchu*.

We catch a further glimpse of Lavoisier's life in 1787 from Arthur Young, who called upon him in October of that year and recorded in his diary:

> To the arsenal, to wait on Monsieur Lavoisier, the celebrated chemist. . . . Dr. Priestley had given me a letter of introduction. I mentioned in the course of conversation his laboratory. . . . That apartment, the operations of which have been rendered so interesting to the philosophical world, I had pleasure in viewing. . . . I was glad to find this gentleman splendidly lodged, and with every appearance of a man of considerable fortune. This ever gives one pleasure: the employments of a state can never be in better hands than of men who thus apply the superfluity of their wealth. From the use that is generally made of money, one would think it the assistance of all others of the least consequence in affecting any business truly useful to mankind, many of the great discoveries that have enlarged the horizon of science having been in this respect the result of means seemingly inadequate to the end: the energetic exertions of ardent minds bursting from obscurity, and breaking the bands inflicted by poverty, perhaps by distress.

Lavoisier's laboratory at the Arsenal, his visit to which Arthur Young records with pleasure, was, indeed, remarkable. Up to that time, there had been nothing to compare with it; and many years were to pass before such a collection of instruments, especially of precision instruments, and chemical apparatus would be put together again as the working tools of a laboratory—probably not until the rise of the modern research institutions. This statement is not exaggerated; for the inventories made by the Revolutionary agents after the confiscation of Lavoisier's property show that his laboratory contained over 13,000 items of glass and other chemical apparatus and specimens, valued at more than 7,000 livres, together with 250 physical instruments. These included his three precision balances, one by Fortin and two by Megnié, valued at 3,500 livres, the historic instruments used in the work of the Commission of Weights and Measures for the determination of the *gram,* the new unit of weight in the metric system, which has since become the international scientific unit of weight.

Most of Lavoisier's life had been passed in Paris. Though he had travelled widely through France, he had only once been outside her frontiers, when he crossed into Switzerland to visit Basle in 1767 on the geological tour with Guettard. In 1786 he had arranged to travel in Switzerland, but the journey had to be postponed and his plans were never realized. He had often expressed an intention to visit England, but had never been able to carry it out, nor was he able to do so in the troubled years that lay ahead.

Although Lavoisier had always been alert to insist on his rights of priority with regard to his scientific discoveries, he appears to have had no similar keenness about his memoirs and reports on agriculture and economics, many of which remained unpublished; others, such as those read in the Provincial Assembly of Orléanais, appeared without the name of the author in the records of the proceedings. It is some evidence of his modesty in this field of activity that he apparently never thought of publishing under his own name a collection of studies that would have given him a high place among the economists of his age. The inspiration for such work seems to have arisen entirely from his desire for the amelioration of the social conditions of the poorest classes of his countrymen.

In scientific matters, however, his attitude had been quite different. He had been criticized for it, not always justly, and some of the criticism has been repeated even in our own time. Before 1787 it had turned on what was regarded as his insufficient recognition of the experiments in which Cavendish had discovered that the burning of inflammable air in common air or in dephlogisticated air produced water. But, as we have already shown, it was Lavoisier who gave the right explanation of the results Cavendish had first achieved.

Two years later there were to be other complaints of this same nature. It was said, and it has been repeated since, that in his *Traité* he did not give to others the credit that was their due, and that he claimed that he was one of those who had discovered oxygen. But the *Traité* was intended, as Lavoisier stated in the preface, as a textbook for students and was a summary of the new system of chemistry that he had devised. It was meant to be a brief presentation of the new system, suitable for beginners. As for his alleged failure to give proper credit to others, this was not the place to expect it. Their work had been repeatedly recognized in memoir after memoir in which he had slowly advanced during these twenty years towards a new system. With regard to the discovery of oxygen, Lavoisier's claim is not unjust. Priestley had isolated the gas and, as we have seen, had called it dephlogisticated air, common air deprived of the phlogiston that, he supposed, was normally present in it by contamination from domestic and other fires and from animal respiration. In Sweden, Scheele, similarly confused with the phlogiston theory, from which he had been unable to break free, had called it "fire air," though Scheele's work was not published until 1777 and therefore not known until Lavoisier's work on the composition of the air was complete. As compared with Priestley, Lavoisier had concluded that common air was composed of two gases, one-fifth being oxygen and the other four-fifths an asphyxiant or *mofette,* later called azote and later still nitrogen; that the oxygen was the part that sustained combustion and respiration and combined with metals on calcination; and that the *mofette,* the nitrogen, was inactive in all these processes.

It would be a curious use of the word "discovery" to deny to Lavoisier any share, to say the least, in the discovery of oxygen. Lavoisier himself had said that there was no principle of scientific conduct that forbade him to give better explanations of other men's discoveries than those they could provide themselves, an attitude to which no man of science could take exception. However, when Fourcroy, his collaborator in 1787, began to speak of the new theory as "the theory of the French chemists," Lavoisier rightly and properly made his views quite clear: "This theory is not, as I hear it called, the theory of the French chemists. It is mine. It is a right that I claim by the judgment of my contemporaries and at the bar of history." History has never questioned the justice of his claim.

XXII

THE REVOLUTION IN CHEMISTRY

LAVOISIER'S THEORY SLOWLY WON ACCEPTANCE FROM HIS contemporaries. For several years Laplace had been convinced of its correctness and had collaborated with Lavoisier in some of his researches. Berthollet accepted it publicly at the Academy of Sciences at a meeting on April 6, 1785; Guyton de Morveau followed in 1786 and Fourcroy in 1787; and later Monge, Meusnier, and Chaptal, adopted the new system. But its most powerful and significant support came from Joseph Black, who was teaching the new chemistry in his lectures in the University of Edinburgh some years before 1784; we do not know the year in which he began to do this, as the only record that survives states the fact in that way. Black's support was more important than that of any other chemist, since he was recognized throughout Europe as the pre-eminent chemical philosopher of his age.

New chemical names (1787)

FROM EARLIEST TIMES MEN HAD GIVEN NAMES TO THE MAterials that they used and the substances that existed around them. In the eighteenth century, many substances retained the names given to them in classical antiquity, or names often coined from their appearance or their properties, or from the place of their discovery or occurrence (Epsom salt), or after their discoverer (fuming liquor of Libavius), or from some superficial resemblance to a familiar substance (butter of arsenic—incidentally, a violent poison), or from the ancient or alchemical association of the metals with the planets (vitriol of Venus—the substance we now call copper sulphate).

Even before the discovery of the composition of water, Lavoisier, Berthollet, and Fourcroy, had been considering with de Morveau, who had inspired the movement, a plan for a new system of chemical names, a new nomenclature in chemistry. Their studies began in 1782, and on April 18, 1787, at a public meeting of the Academy, Lavoisier presented an account of the first results of their work, the object of which was the formulation of a scientific nomenclature to replace the ancient system that had grown up almost haphazardly through long centuries, during which little or nothing had been known of the chemical composition of substances. Under the new system, the name given to a substance was designed to indicate its chemical composition.

That Lavoisier was chosen by his collaborators to read the first memoir relating to their joint work and to explain the new principles on which they had devised an improved system of chemical names shows, not only the standing that he had attained among contemporary

French chemists, but also the recognition that his new theory had gained.

Lavoisier explained that, although there had been earlier attempts to classify similar substances under distinctive names, such as applying the term "vitriols" to the salts obtained by the solution of metals in vitriolic acid and "niters" to those containing "nitrous acid," such reform was limited. The problem was difficult, because the science of chemistry was changing rapidly with new discoveries. De Morveau would present details of the new nomenclature in a second memoir, while he, Lavoisier, would expound the principles on which the proposed reforms were based. Tables showing the new system and its details would be exhibited in the Academy for discussion and criticism.

Said Lavoisier:

> Languages are not only intended, as is commonly supposed, to express ideas and images by signs, but also are real analytical systems by means of which we advance from the known to the unknown, and to a certain extent in the manner of mathematicians: let us try to expound this idea.
>
> Algebra is the analytical method *par excellence:* it has been contrived to facilitate the operations of the understanding, to make reasoning more concise, and to contract into a few lines what would have needed many pages of discussion; in short, to lead by a more convenient, rapid, and unerring method, to the solution of the most complicated questions. But a moment's reflection readily shows us that algebra is a real language: like all languages it has its representative signs, its methods, its grammar, if I may use this expression. Thus an analytical method is a language and a language is an analytical method; and these two expressions are, in some sense, synonymous.

Lavoisier then referred to the *Logic* (1780) of the philosopher de Condillac, whose works had greatly influenced him:

> This truth has been expounded with infinite exactness and clarity in the *Logic* of the Abbé de Condillac, a work which can never be too much studied by young men who are destined for the sciences, and from which we cannot avoid borrowing a few ideas. He showed in that work how the language of algebra could be translated into common speech and *vice versa;* how the progress of the understanding is the same in both; and how the art of reasoning is the art of analyzing.
>
> If languages really are instruments fashioned by men to make thinking easier, they should be of the best possible kind; and to strive to perfect them is indeed to work for the advancement of science. For those who are beginning the study of a science, the perfecting of its language is of high importance. We shall be convinced of this if we reflect for a moment on the manner in which we acquire our knowledge.
>
> In our early infancy our ideas arise from our needs; the sensations brought about by our needs produce the idea of objects fit to satisfy those needs, and unconsciously by a train of sensations, of observations, and of analyses, there is formed a successive and connected series of ideas, of which an attentive observer may to some extent discover the thread and the connection, and which constitutes the totality of what we know.
>
> When we first begin to study a science, we are very much like children, so far as that science is concerned; we follow the same path in forming our ideas as Nature does in forming theirs. Just as with a child, the idea is a sequel or an effect of the sensation, and it is the sensation that gives birth to the idea. The ideas of

a beginner in the study of the physical sciences should likewise be only the immediate effect of an experiment or observation.

But the beginner in science, said Lavoisier, is rather worse off than the child; for if the child finds himself deceived by the salutary or harmful effects of the objects surrounding him, Nature gives him many means of remedying his mistakes, since experience constantly corrects his judgments. False judgment leads to pain and privation; correct judgment leads to enjoyment and pleasure. Pain and pleasure are good instructors and they encourage exact reasoning.

But in science the consequences are different; false judgments do not affect our existence or our happiness, and no bodily interest compels us to rectify our mistakes. On the contrary, our imagination continually tends to take us beyond the limits of truth and our self-confidence induces us to draw conclusions that are not immediately derived from the facts.

"It is therefore not astonishing," added Lavoisier, "that, in the early childhood of chemistry, suppositions instead of conclusions were drawn, that those suppositions transmitted from age to age were changed into presumptions, and that these presumptions were adopted and regarded as fundamental truths even by the ablest minds."

The only way to prevent this is to simplify our reasoning, Lavoisier stated, and to control it by constant subjection to the test of experiment, to accept only those facts that are Nature's own proofs and cannot deceive us, and to seek for truth only in a connected series of experiments and observations, made in the same way as mathematicians proceed to the solution of a problem by the simple arrangement of the data and by such simple reasoning and precise conclusions that they never lose sight of the evidence that is their guide. "This method, which must be introduced into the study and teaching of chemistry, is closely connected with the reform of its nomenclature. A well-composed language adapted to the natural and successive order of ideas will bring in its train a necessary and immediate revolution in the method of teaching and will not allow teachers of chemistry to deviate from the course of Nature; either they must reject the nomenclature or they must irresistibly follow the course marked out by it. The logic of the sciences is thus essentially dependent on their language."

Proceeded Lavoisier:

> If, after having considered languages as analytical methods, we consider them merely as a collection of representative signs, they will afford us some observations of another kind. In this second aspect, we shall have three things to distinguish in every physical science: the series of facts that constitute the science; the ideas that call the facts to mind; and the words that express them. The word should give birth to the idea; the idea should depict the fact; they are three impressions of one and the same seal; and as it is the words that preserve and transmit the ideas, it follows that the science can never be brought to perfection, if the language be not first perfected, and that however true the facts may be and however correct the ideas to which they give rise, they will still transmit only false impressions, if there are no exact expressions to convey them. The perfecting of the nomenclature of chemistry, considered from this point of view, consists in conveying the ideas and facts in their strict verity, without suppressing anything that they present, and

> above all without adding anything to them; it should be nothing less than a faithful mirror; for we can never too often repeat that it is not Nature, nor the facts that Nature presents, but our own reasoning that deceives us.

So far, he said, these principles had not been applied to the language of chemistry; nor could they have been applied in times when experiment was not yet known, when all was given to imagination and almost nothing to observation, when even the method of studying the science was unknown.

Lavoisier added:

> Further, some of the expressions used in chemistry were introduced by the alchemists; it would have been difficult for them to convey to their readers what they did not have themselves, just and correct ideas. Moreover, their aim was not always to make themselves understood. They made use of an enigmatical language, which was peculiar to themselves, which most frequently had one meaning for the adepts and another for the uninitiated, and which had nothing exact or intelligible for either the one or the other. So it is that the oil, the mercury, and even the water, of these philosophers were neither oil, nor mercury, nor water, in the sense in which we use these words. . . . Another class of philosophers, who have not much less disfigured the language of chemistry, are the systematical chemists. They have erased from the assembly of facts those that would not square with their ideas; they have somehow changed those that they wished to preserve; and they have accompanied them with an apparatus of argument that makes us lose sight of the facts themselves in such a way that the science in their hands is no more than a fabric woven by their imagination.

For the first time, after twenty years of experimental research in the culmination of which he had revolutionized the science of chemistry, Lavoisier was expounding his philosophy of science. In this illuminating memoir he was dealing with what is a fundamental instrument to all sciences, an exact system of analysis or a language, and with one that was to be adapted to the present state and future development of the new chemistry of which he was the supreme architect.

"It is now time," he urged, "to rid chemistry of every kind of impediment that delays its advance; to introduce into it a true spirit of analysis; and we have sufficiently demonstrated that it is by the perfecting of its language that this reform must be brought to pass. We are doubtless very far from knowing the whole and all the parts of this science: it is therefore to be expected that a new nomenclature, although formed with all possible care, must be far from perfect; but provided that it has been undertaken upon sound principles, provided that it is a method of naming rather than a nomenclature, it will naturally adapt itself to future discoveries; it will indicate beforehand the place and name of new substances that may be discovered, and it will need no more than particular amendment in some details."

Great discussions about the constituent principles of bodies and their elementary particles, said Lavoisier, would be in contradiction with the opinions that he had so far set out. But he and his collaborators had made a most significant and essential decision; in effect, they had decided to apply the definition of the chemical element that Boyle had formulated in 1661 and had failed to apply: *We shall content ourselves here with regarding as simple all the substances that we cannot decompose, all that we obtain in the last resort by chemical analysis.*

With these historic words Lavoisier stripped from chemistry the last shreds of its medieval heritage.

Lavoisier, however, gave full recognition to the possibility that the substances that then appeared to be simple might later prove to be compound. The simple substances, "doubtless improperly called so," as Lavoisier insisted, were the first to be named in the new system. Most of them already had well-known names and the reformers proposed to retain these unless there was some strong reason to the contrary, as when they led to false ideas. In those cases new names were to be given, generally derived from Greek and expressing the most general and characteristic properties of the substances designated.

For the application of these principles we have to turn to the second memoir in the series, read by de Morveau at the Academy on May 2, 1787. Here the substances that had not been decomposed were divided into five classes. The first class included light and heat, oxygen and hydrogen, the last so-called because it was one of the constituents of water (*ὕδωρ*, water: *γείνομαι*, I produce). The second class contained the radical principles of the acids, such as sulphur and phosphorus, and also carbon, as the principle of charcoal was now named, and azote (*ἀ*, not: *ζωή*, life), the irrespirable gas of the atmosphere (later renamed nitrogen), as well as the unknown radicals of a large number of acids. The third class comprised the metals, arsenic, molybdenum, tungsten, manganese, nickel, cobalt, bismuth, antimony, zinc, iron, tin, lead, copper, mercury, silver, platinum, and gold. The metals were thus at last regarded as simple, whereas the phlogiston theory had taken them to be compounds. This was a revolutionary change. The fifth class consisted of the earths, silica, alumina, baryta, lime, and magnesia; and the sixth of the alkalis, potash, soda, and ammonia.

The list drawn up by the reformers and expounded by de Morveau included fifty-five "substances not decomposed." Although this might, in some respects, be regarded as the first list of chemical elements in the modern sense, the chemists who had collaborated in its production seem to have been expressing themselves tentatively and with diffidence; although they used the term "simple substances," they always took care to explain that this meant "undecomposed substances." Not until the publication of Lavoisier's *Traité* in 1789 did a table of elements appear, and in that table the number was considerably reduced.

After dealing with the undecomposed substances, Lavoisier, in the opening memoir of April 18, 1787, explained that for bodies composed of two simple substances classification was necessary, as they were very numerous. The acids formed one of these classes. One of the two simple substances of which every acid was composed, namely oxygen, the acidifying principle, was common to them all, and the other varied from acid to acid; it was from the latter that the specific name was to be formed. For the application of this principle, we turn again to de Morveau. The acid produced by the combination of sulphur with oxygen was renamed *sulphuric acid,* but to distinguish the similar acid containing less oxygen, the latter was named *sulphurous acid.* Their salts became *sulphates* and *sulphites* respectively. In a similar way, the acid formed by the combination of oxygen with carbon was named *carbonic acid* and its salts became *carbonates.* The acids of phosphorus were named *phosphoric acid* and *phosphorous acid,* analogous to *sulphuric* and *sulphurous* acids. "Nitrous acid" was renamed *nitric acid,* and the name *nitrous acid* was given to the similar acid containing less oxygen; and their salts were named *nitrates* and *nitrites.*

In the same way, Lavoisier pointed out, the calces of metals contained a principle common to them all and a particular principle proper to each. The class comprising the metallic calces must be given a general or generic name, just as the class of acids had been so named, while each calx must have a specific name, the name of the metal from which it was formed. In de Morveau's memoir, it was accordingly explained that the metals retained their names, but their calces, now shown to be compounds with oxygen, were named *oxides*. This was the general or class name, a particular oxide being given the specific name of its metal, e.g., the compound of zinc and oxygen became *zinc oxide*. Likewise, the salts were given names that included the metal and the acid from which they had been formed, e.g., *sulphate of copper, nitrate of lead,* as Lavoisier explained, when he turned to consider the great number of bodies composed of three simple substances. "We have had to consider," he said, "in the bodies that form this class, such as, for example, the neutral salts, firstly, the acidifying principle that is common to all, secondly, the acidifiable principle that constitutes their particular acid and, thirdly, the saline, earthy, or metallic, base that determines the particular kind of salt. We have taken the name of each class of salt from that of the acidifiable principle common to all the members of that class and we have then distinguished each kind by the name of its particular saline, earthy, or metallic base."

In conclusion, Lavoisier added:

> It may be thought that we could not have accomplished these different aims without doing frequent injury to established custom and adopting denominations which at first appear harsh and barbarous; but we have observed that the ear readily becomes accus-

> tomed to new words, especially when they are connected together in a general and rational system. Besides, the names that are now in use, such as those of *powder of algaroth, salt of alembroth, pompholix, phagedenic water, turbith mineral, æthiops, colcothar,* and many others, are neither less harsh nor less extraordinary; much practice and a great memory are necessary to remember the substances that they represent, and especially to recollect to what genus of combination they belong. The names of *oil of tartar per deliquium, oil of vitriol, butter of arsenic, butter of antimony, flowers of zinc,* &c., are still more ridiculous, because they give rise to false ideas; because, properly speaking, there does not exist in the mineral kingdom, and especially in the metallic, either butter or oil or flowers; in short, because the substances designated by these deceitful names are for the most part violent poisons.
>
> Shall we be forgiven for having changed the language that our masters have spoken, made illustrious and passed on to us? We hope so, and so much the more because it was Bergman and Macquer who urged this reform. The learned professor of Upsala, M. Bergman, wrote to M. de Morveau not long before his death: *Have no mercy on any improper denomination. Those who already know will always understand: those who do not as yet know it will understand the sooner.* Called upon to cultivate the field that has produced such abundant harvests for these chemists, we have held it to be our duty to fulfil the last wish that they expressed.

The new nomenclature met with general acceptance, although its welcome at the Academy was not enthusiastic. The memoirs of Lavoisier and de Morveau, together with a memoir by Fourcroy explaining the tables accompanying the memoirs, a synonymy of the old and new names, a dictionary of the new names, and a memoir by Hassenfratz and Adet on new chemical characters or symbols, were published in one volume at Paris in 1787 under the title of *Méthode de Nomenclature Chimique,* and an English version, *Method of Chymical Nomenclature,* translated by James St. John, was published in London in 1788. From this time on the chemical name of a substance indicated the elements it was composed of and the manner in which they were combined. The name described the chemical composition.

The simplicity and logicality of the new system made a great appeal to chemists. It remains only to say that it is essentially the same as that still in use in chemistry and that it has fulfilled that most difficult aim set by Lavoisier, to be such as would be adapted to future discovery.

The Elements of Chemistry (1789)

TWO YEARS AFTER THE APPEARANCE OF THE NEW NOMENCLATURE, Lavoisier published his *Traité Élémentaire de Chimie* in Paris in 1789. Other editions followed in rapid succession. The book was well received. An English translation by Robert Kerr under the title *Elements of Chemistry* appeared in Edinburgh in 1790, a reproduction of the title-page of which is shown in Fig. 21. Four further editions appeared in Edinburgh, the fifth and last in 1802. German, Italian, Spanish and Dutch translations were published and republished, and there were American editions. All have become rare. Lavoisier's original manuscript is preserved in the archives of the Academy of Sciences in Paris.

The *Traité* or *Elements* aimed at presenting a complete and detailed account of the new system of chemistry and included extensive descriptions of experiments and apparatus. It laid the foundations of modern chemistry as New-

ELEMENTS

OF

CHEMISTRY,

IN A

NEW SYSTEMATIC ORDER,

CONTAINING ALL THE

MODERN DISCOVERIES.

ILLUSTRATED WITH THIRTEEN COPPERPLATES.

BY MR LAVOISIER,

Member of the Academy of Sciences, Royal Society of Medicine, and Agricultural Society of Paris, of the Royal Society of London, and Philoſophical Societies of Orleans, Bologna, Baſil, Philadelphia, Haerlem, Mancheſter, &c. &c.

TRANSLATED FROM THE FRENCH,

BY ROBERT KERR, F.R. & A.SS.E.

Member of the Royal College of Surgeons, and Surgeon to the Orphan Hoſpital of Edinburgh.

EDINBURGH:

PRINTED FOR WILLIAM CREECH, AND SOLD IN LONDON BY G. G. AND J. J. ROBINSONS.

MDCCXC.

21. Title-page of *The Elements of Chemistry* (first English edition, 1790).

ton's *Principia* had laid the foundations of modern physics a century earlier.

Lavoisier explained in the Preface that he had begun to write with the object merely of extending and explaining his previous work of 1787 on the renaming of chemical substances. Exposition of the principles of that system gradually developed, however, until he found that he was engaged, without being able to prevent it, upon the task of writing a complete treatise on chemistry. As the earlier work had shown, the language of the science and the science itself were almost one and the same thing.

Much of what he had said in 1787 Lavoisier repeated in the Preface to the *Traité*. The importance of fact and experiment was again emphasized: "Convinced of these truths, I have imposed upon myself the law of never advancing but from the known to the unknown, of deducing no consequence that is not immediately derived from experiments and observations." Accordingly, he would have nothing to say about the ultimate constituent and elementary parts of matter. The attempts to reduce all bodies in nature to three or four elements arose from ancient ideas that had come down from the Greek philosophers. The idea of four elements was a mere hypothesis formulated before there was any understanding of the principles of experimental science.

He added:

> I shall therefore content myself with saying that, if by the term *elements,* we mean to express the simple and indivisible molecules that compose bodies, it is probable that we know nothing about them: but if, on the contrary, we express by the term *elements* or *principles of bodies* the idea of the last point reached by analysis, all substances that we have not yet been able to decompose by any means are elements to us;

> not that we can assert that these bodies that we consider as simple are not themselves composed of two or even a greater number of principles, but, since these principles are not separated, or rather since we have no means of separating them, they are to us as simple substances, and we must not suppose them compounded until experiment and observation have proved them to be so.

The *Traité* was arranged in three parts. The first comprised what was essentially the new system of chemistry set up by Lavoisier. The second dealt with the acids and bases and the formation of salts, and the third with the methods and apparatus of chemistry.

In the first part, Lavoisier explained his views on heat and the nature of aeriform fluids, now renamed gases. A gas was produced by the combination of caloric, the matter or material cause of the sensation of heat, with the base of the gas. Oxygen gas, for example, was formed from oxygen base and caloric. Heat and light, frequently the concomitants of chemical change, were produced in the combustion of bodies by release of this material heat or caloric. The formation and composition of the atmosphere was explained in terms of this theory, and then came an account of the analysis of the common air of the atmosphere into two gases, one respirable and the other not. At this point Lavoisier gave, for the first time, full details of his classic experiment on the composition of air and a diagram (Fig. 22) of the apparatus. This is one of the most famous of all experiments in chemistry and, indeed, in the long history of science. We quote Lavoisier's own account of it:

> I took a matrass A of capacity about 36 cubic inches with a long neck BCDE of internal diameter about 6

to 7 lines. I bent the neck as shown, in such a way that it could be placed on a furnace MMNN, while the extremity E of its neck could be inserted under the bell-jar FG, placed in a trough of mercury RRSS. I put 4 ounces of pure mercury into the matrass and then, by suction with a siphon which I introduced under the bell-jar FG, I raised the mercury to LL: I carefully

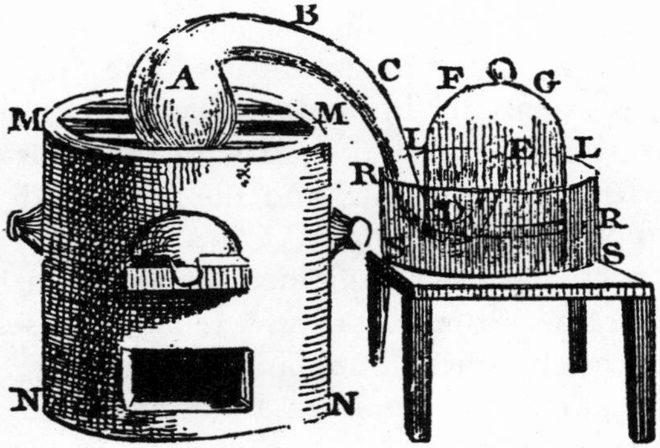

22. The analysis of air.

marked this height with a strip of gummed paper and I noted accurately the readings of the barometer and thermometer.

The apparatus being thus prepared, I lighted a fire in the furnace MMNN, which I kept up almost continually for twelve days, in such a way that the mercury was heated nearly to the degree necessary to make it boil.

Nothing remarkable occurred during the first day: the mercury, though not boiling, was in a state of continual evaporation; it covered the inside of the vessels with small drops, which, at first very minute, afterwards increased in size and, when they had acquired a certain volume, fell back into the bottom of the vessel and were re-united with the rest of the mercury. On the second day, I began to see swimming on the surface of the mercury small red particles, which gradually increased in number and size during four or five days, after which they ceased to increase and remained completely unchanged. At the end of twelve days, seeing that the calcination of the mercury made no further progress, I extinguished the fire and allowed the vessel to cool. The volume of air contained in the matrass, in its neck and in the bell-jar, reduced to a pressure of 28 inches and 10° [R.]* of the thermometer, was about 50 cubic inches at the beginning of the experiment. When the experiment was completed, the volume of the air, reduced to the same pressure and temperature, was only about 42 to 43 cubic inches; a decrease in volume of about one-sixth had therefore occurred. On the other hand, having carefully collected the red particles that had been formed and having separated them as much as possible from the liquid mercury with which they were mixed, I found that their weight was 45 grains.

I was obliged to repeat this calcination of mercury in closed vessels several times, because it is difficult in one and the same experiment to recover all the air experimented with and all the red particles or calx of mercury produced. It will often happen that I shall thus combine in one statement the result of two or three experiments of the same kind.

The air that remained after the experiment, and which had been reduced to five-sixths of its volume by the calcination of the mercury, was no longer fit either for respiration or for combustion; for animals that were put into it perished in a few seconds and candles were extinguished immediately, as if they had been plunged into water.

* The Réaumur thermometric scale, on which ice melts at 0° and water boils at 80°.

On the other hand, I took the 45 grains of red matter that had been formed during the experiment; I put them into a very small glass retort, to which was fitted an apparatus suitable for receiving the liquid or aeriform products that might be separated: having lighted the fire in the furnace, I observed that, in proportion as the red matter grew hot, the intensity of its color increased. When the retort was almost glowing hot, the red matter began gradually to decrease in bulk and in a few minutes it disappeared completely; at the same time 41½ grains of liquid mercury were condensed in the small receiver and 7 to 8 cubic inches of an elastic fluid, much fitter than the air of the atmosphere for supporting both combustion and the respiration of animals, passed over into the bell-jar.

A part of this air being put into a glass tube of about 1 inch diameter, a candle was plunged into it and burned with a dazzling splendor; charcoal, instead of being consumed quietly as in ordinary air, burned in it with a flame and a kind of decrepitation, like phosphorus, and with a brilliancy of light that the eyes could hardly bear. . . .

In reflecting upon the circumstances of this experiment, we see that the mercury during its calcination absorbs the salubrious and respirable part of the air . . . ; that the part of the air that remains is a kind of *mofette,* incapable of supporting combustion and respiration; and therefore that the air of the atmosphere is composed of two elastic fluids of a differing and, if I may say so, opposite nature.

A proof of this important truth is that, by recombining the two elastic fluids that we have thus obtained separately, namely, the 42 cubic inches of *mofette,* or irrespirable air, and the 8 cubic inches of respirable air, we reproduce an air similar in every respect to the air of the atmosphere and fit very nearly to the same degree for combustion, the calcination of metals, and the respiration of animals.

These experimental details were the same as those given in the memoir read to the Academy on May 3, 1777, on the respiration of animals. In the *Traité,* however, a diagram of the apparatus was given for the first time. In this experiment, Lavoisier pointed out, the composition of the air had been demonstrated both by analysis and synthesis.

The nomenclature for the constituents of atmospheric air was then discussed. The term *gas,* first used by van Helmont and recently re-introduced by Macquer, was now accepted by Lavoisier for the kind of substance previously known as an *elastic fluid.* The terms *oxygen* and *azote* were once more explained.

The experiments quoted on the combustion of phosphorus in oxygen are similar to those that we have already described. Those in which sulphur was used had given no quantitative results of any accuracy. Lavoisier had been able to show that sulphur gained in weight on combustion and that the gain in weight was equal to the weight of oxygen absorbed, but he had, he said, been unable to determine the relative proportions of sulphur and oxygen entering into the composition of the acid produced. It is interesting to note, however, that Lavoisier's laboratory notebook has an entry for May 18, 1785, recording that the combining proportions of sulphur and "vital air" were as 38.98 to 61.02, which is roughly as 39 to 61. As we now know that the proportions for sulphur trioxide are as 40 to 60, this was far from being an unsatisfactory result but, of course, other experiments may have given different values and Lavoisier may have been led to doubt its accuracy.

The system of nomenclature for the acids was again

detailed and the term *oxide* was explained as before. Various experiments to demonstrate the composition of water, both by analysis and synthesis, were described without any reference to the disputes that had arisen, but with the comment: "From these experiments, both analytical and synthetic, we may now affirm that we have ascertained, with as much certainty as is possible in physical or chemical subjects, that water is not a simple elementary substance, but is composed of two elements, oxygen and hydrogen." Values were given from experiments on the decomposition and synthesis of water to show that 100 parts by weight of water were composed of 85 parts by weight of oxygen and 15 parts by weight of hydrogen. (The value now accepted is 89 of oxygen to 11 of hydrogen.)

The vegetable and animal acids, said Lavoisier, were oxides, just as the mineral acids were. The vegetable acids had compound groupings of hydrogen and carbon, and it was the union of these compound groups or radicals, as they were later called, with oxygen that produced these acids. This conception of organic acids as containing compound radicals of carbon and hydrogen was to prove useful later on in the nineteenth century in the development of organic chemistry.

Lavoisier next dealt with vinous fermentation in one of the most interesting sections of the book. The fermentation of fruit juices or of any saccharine matter was accompanied by the production of carbonic acid gas and spirit of wine, to which Lavoisier now restored its old Arabic name of *alcohol.*

He wrote:

> This operation is one of the most striking and extraordinary of all those that chemistry presents to us;

> and we must examine whence proceed the disengaged carbonic acid gas and the inflammable spirit produced, and how a sweet body, a vegetable oxide, can thus be converted into two such different substances, whereof one is combustible and the other is highly incombustible. It will be seen that, to solve these two questions, we must first know the analysis of the fermentable body and the products of the fermentation; for nothing is created in the operations either of art or of Nature, and it can be taken as an axiom that in every operation an equal quantity of matter exists both before and after the operation, that the quality and quantity of the principles remain the same, and that only changes and modifications occur. The whole art of making experiments in chemistry is founded on this: we must always suppose an exact equality between the principles of the bodies examined and those obtained by analysis. Thus, since must of grapes gives carbonic acid gas and alcohol, I can say that *must of grapes = carbonic acid + alcohol.* Thence it follows that we can ascertain in two ways what happens in the vinous fermentation, first, by determining the nature and the principles of the fermentable body, and, secondly, by carefully observing the products of the fermentation, and it is evident that the knowledge that we can acquire about one of these leads to accurate conclusions about the nature of the other.

This was the first time that anyone had asserted the fundamental principle that the Law of the Conservation of Mass or the Law of the Indestructibility of Matter applied to chemical change. It had been implicit in the researches of Black and of Cavendish, but it had never been previously formulated:

> *Nothing is created in the operations either of art or of Nature, and it can be taken as an axiom that in every*

operation an equal quantity of matter exists both before and after the operation.

Lavoisier then described his experiments on the vinous fermentation. He determined the amounts of hydrogen, oxygen, and carbon in sugar. He weighed a quantity of sugar, added yeast and water in known amounts, and allowed the mixture to ferment. He determined the weights of carbonic acid gas and water that were given off during fermentation and he weighed the residual liquor, the components of which were then separated and analyzed to determine their elementary composition. In a series of five tables Lavoisier showed that the total weight of the final residue and the products was equal to the total weight of the materials originally taken. Moreover, he showed also that the total weight of each constituent element before and after the chemical change remained the same. These tables constitute the first chemical "balance sheet" and illustrated, for the first time, the principle that in chemical reaction there is no loss of matter, but only change in its forms of aggregation.

The principle that Lavoisier was here demonstrating leads also to the chemical equation, as he himself saw and pointed out: "We may consider the substances submitted to fermentation, and the products resulting from that operation, as forming an algebraic equation; and, by successively supposing each of the elements in this equation unknown, we can calculate their values in succession, and thus verify our experiments by calculation, and our calculation by experiment reciprocally. I have often successfully employed this method for correcting the first results of my experiments, and to direct me in the proper road for repeating them to advantage."

But the most revolutionary feature of the *Traité* was the

TABLE OF SIMPLE SUBSTANCES.

Simple ſubſtances belonging to all the kingdoms of nature, which may be conſidered as the elements of bodies.

New Names.	*Correſpondent old Names.*
Light	Light.
Caloric	Heat. Principle or element of heat. Fire. Igneous fluid. Matter of fire and of heat.
Oxygen	Dephlogiſticated air. Empyreal air. Vital air, or Baſe of vital air.
Azote	Phlogiſticated air or gas. Mephitis, or its baſe.
Hydrogen	Inflammable air or gas, or the baſe of inflammable air.

Oxydable and Acidifiable ſimple Subſtances not Metallic.

New Names.	*Correſpondent old names.*
Sulphur Phoſphorus Charcoal	The ſame names.
Muriatic radical Fluoric radical Boracic radical	Still unknown.

Oxydable and Acidifiable ſimple Metallic Bodies.

New Names.	*Correſpondent Old Names.*
Antimony	Regulus of Antimony.
Arſenic	Regulus of Arſenic.
Biſmuth	Regulus of Biſmuth.
Cobalt	Regulus of Cobalt.
Copper	Regulus of Copper.
Gold	Regulus of Gold.
Iron	Regulus of Iron.
Lead	Regulus of Lead.
Manganeſe	Regulus of Manganeſe.
Mercury	Regulus of Mercury.
Molybdena	Regulus of Molybdena.
Nickel	Regulus of Nickel.
Platina	Regulus of Platina.
Silver	Regulus of Silver.
Tin	Regulus of Tin.
Tungſtein	Regulus of Tungſtein.
Zinc	Regulus of Zinc.

Salifiable ſimple Earthy Subſtances.

New Names.	*Correſpondent old Names.*
Lime	Chalk, calcareous earth. Quicklime.
Magneſia	Magneſia, baſe of Epſom ſalt Calcined or cauſtic magneſia
Barytes	Barytes, or heavy earth.
Argill	Clay, earth of alum.
Silex	Siliceous or vitrifiable earth.

23. The first table of the chemical elements (1789).

list of chemical elements given in it. We have seen that the *New Chemical Nomenclature* gave a list of "substances not decomposed." Here, however, Lavoisier calls them "simple substances belonging to all the kingdoms of nature, which may be considered as the elements of bodies." The list of elements is much shorter than the list of "substances not decomposed" given two years earlier in 1787 in the memoirs on the new nomenclature, for Lavoisier had removed from the earlier list the unknown organic radicals, which he had since come to regard as composed of hydrogen and carbon. We reproduce here (Fig. 23) this first table of chemical elements from the first edition (1790) of the English version of the *Traité*.

In commenting on the table of elements, Lavoisier wrote:

> Chemistry in subjecting to experiments the various bodies in Nature aims at decomposing them so as to be able *to examine separately the different substances that enter into their composition*. In our time this science has made very rapid progress, as will be easily seen by consulting the various writers on chemical systems. Oil and salt were formerly regarded as elements of bodies, whereas, experiment and observation having brought new knowledge, it has since been shown that the salts are not simple but composed of an acid and a base, and that their neutrality results from this combination. Modern discoveries have further extended by several stages the limits of analysis: they have enlightened us on the formation of acids and have shown us that acids are formed by the combination of an acidifying principle common to all, namely, oxygen, and of a radical that is particular to each acid. . . . Chemistry advances towards its goal and its perfection by dividing, subdividing, and resubdividing; and we do not know what the limit of

> its achievements may be. Therefore we cannot assert that what we at present suppose simple is so in fact; all that we can say is that such a substance is the actual limit that chemical analysis has attained and that it is not further subdivisible in the present state of our knowledge. We may suppose that the earths will soon cease to be counted among the number of simple substances; they are the only bodies in the whole of this class that have no tendency to combine with oxygen and I am much inclined to think that this indifference towards oxygen, if I might be allowed to use this expression, results from their being already saturated with it. The earths, according to this view, should be compound substances, perhaps metallic oxides oxygenated up to a certain point. At the most this is only a mere conjecture. I hope that the reader will take care not to confound what I relate as truths of fact and experience with what is yet only hypothetical. I have omitted the fixed alkalis, potash and soda, from this table because these substances are evidently compound, although we do not know as yet the nature of the principles that enter into their composition.

In this omission Lavoisier was correct; shortly after the turn of the century Davy decomposed the alkalis, potash and soda, and obtained their elements, potassium and sodium.

The *Traité* was, as Lavoisier said, written for those beginning chemistry; but the new chemistry was so different from what had gone before, from what it swept away, that there were few who were not beginners. This treatise marks the real foundation of modern chemistry. We have, for good reasons, abandoned Lavoisier's generalization that all acids contain oxygen and we have found that the energy changes that occur in chemical reactions are not so simple as he supposed them to be, but his principles are those on which our modern science of chemistry has been erected.

With the publication of the *Traité*, the revolution that Lavoisier had seen to be possible as far back as 1773 was at last achieved and by his own tireless effort. As he himself wrote in 1791: "All young chemists adopt the theory and from that I conclude that the revolution in chemistry has come to pass."

XXIII

STATES-GENERAL AND REVOLUTION (1789)

BUT THE TIDE OF ANOTHER REVOLUTION WAS NOW RISING fast about Lavoisier. The financial problems of France had become hopelessly insoluble, and her people were groaning more loudly and becoming more restive under oppressive taxes and social inequalities. The collapse of Calonne's magic methods of raising money brought the Government face to face with bankruptcy; and Calonne was dismissed on April 8, 1787, to be succeeded by Loménie de Brienne, who presented certain reforms. A land-tax and a stamp-tax were submitted to the Parliament of Paris for registration on July 6, 1787, registration being one of the Parliament's hereditary privileges and a necessary legal formality. The Parliament refused and demanded a statement of expenditure, but, after a month of debate and argument, they were commanded to appear at Versailles on August 6 and ordered to register. Assembled the next day in Paris, they "registered" that their action of the previous day was null and void, that by error the practice had been wrong through the centuries, and that the only authority competent to act was the States-General.

On August 15, by *lettres de cachet,* they were exiled to Troyes, to return thence to Paris on September 20 amid public rejoicings, after an agreement with the Minister that there would be neither land-tax nor stamp-tax nor mention of States-General, but an edict for a new kind of loan and another for Protestant emancipation. However, when the King demanded the registration of these edicts, he was met with appeals to summon the States-General forthwith. Resistance and more *lettres de cachet* and more wrangling followed. The temporary popularity of the Parliament in the public mind was eclipsed in obloquy when they expressed the view that the States-General should meet in the same form as on the last previous assembly in 1614, whereby the Parliament revealed that reform was not their intent.

Powerless to mend matters, as the months wore on into 1788, the King on August 8 of that year as a last resort summoned the States-General to assemble in May, 1789. Centralized government had almost broken down; and the King's motive was clear.

The great changes that were to come in this most momentous of all revolutions, many of which deeply influenced the future political life and development of other nations, were all effected in Paris. Many of those who played their parts in these stirring events were personally known to Lavoisier, many were his friends and associates, in the Government, in the Academy, and in the wider intellectual life of eighteenth-century France, and many of them had sought his advice and counsel on scientific, administrative, social, and economic problems.

The States-General assembled on May 5, 1789. On June 17 the Third Estate adopted the title of National

Assembly and assumed sovereign powers. Six days later the three orders assembled together and an ominous National Guard was organized. Constitutional monarchy breathed a while, but somewhat insecurely, while the Commune of Paris gained the real control of France. The National Assembly became the Legislative Assembly on October 1, 1791, and finally on September 21, 1792, the National Convention, which declared France a republic. Louis XVI was guillotined on January 21, 1793, and Marie Antoinette on October 16 of that year. But we retrace our steps to follow Lavoisier through these troubled years that witnessed the birth-agony of a new France.

The King's command of August 8, 1788, that the States-General should assemble in May, 1789, was received with enthusiasm throughout France. To his colleagues on the Intermediary Commission of the Provincial Assembly of Orléanais, Lavoisier wrote to say that he now felt that matters were coming to a head and added: "If claims are raised, principally in Court circles, in opposition to public opinion, they are strongly dictated by interest and prejudice. Today the nation is too enlightened not to know that its duty is to act in the interests of the majority, that, if exceptions are to be allowed in favor of any class of citizens, particularly with regard to taxes, they can be made only in favor of the poor, and that inequality of taxation cannot be tolerated except at the expense of the rich." He hoped to see the plans that he had proposed to the Committee of Agriculture and the Provincial Assembly of Orléanais put into operation, together with a decision for the uniform taxation of all classes.

But the summons assembling the States-General had prescribed neither the method of election nor the number of representatives for each order. The antiquated Parliament of Paris thought it proper that the Third Estate should be content with the humiliating position that they had occupied when the States-General had previously met 175 years earlier, while the Third Estate claimed that they should have as many representatives as Nobles and Clergy combined. Pamphleteers were busy on both sides; an incomplete collection for the last months of 1788 contains 2,500 of these productions. Lavoisier himself put before the Minister a scheme of election based on an earlier plan submitted to the King by Turgot in 1774.

After briefly insisting that the States-General must be truly representative of the nation and must be so organized as to achieve, so far as possible, the greatest good for all whom it represented, the memoir, as was Lavoisier's custom, opened with a detailed historical account of how Frenchmen had governed themselves and how the States-General had come into being centuries earlier. It proposed that the provinces be divided into departments, departments into districts, and districts into municipalities, for the purpose of the new election. All municipalities would elect representatives to the district assemblies, the district assemblies would elect representatives to the Provincial Assemblies, and these, in turn, would elect representatives to the States-General, which would therefore be truly representative of the nation. The qualifications of those who were eligible to vote as first electors, the electors in the municipalities, were not specified; but it is probable that Lavoisier had in mind some such qualification as Turgot had suggested, a qualification based on property.

These proposals had two considerable advantages to recommend them: they would solve the problem of the double representation that the Third Estate were claiming, and they envisaged the elected members as sitting in one assembly, not in "houses" of three separate orders, by which one order might reject the resolutions of the other. Moreover, the two innovations advocated by Lavoisier had already been successfully put into operation in the Provincial Assemblies, where, however, the members had in the first place been chosen by the King.

Lavoisier here made clear his ideas of the principles of constitutional government:

> Let us speak frankly: legislative power does not reside in the King alone, but in the concurrence of his will and that of the nation. The King and his Ministers have recognized this principle with regard to taxation; they have agreed that no subsidy or subvention can be raised unless it is agreed to and authorized; and, indeed, the contrary principle would attack the inviolable right of property. And what then? The King, as he himself declares, cannot make a law to dispose of the meanest possessions of his subjects, and yet he would be able to make laws to dispose at his will of their liberty, their honor, their lives! What good is there in respecting the right of property, if other rights, no less inviolable but much more important, are to be trampled underfoot? Let us therefore take it as established that, whether a law must be proposed by the King and approved by the people, or whether it must be proposed by the people and approved by the King, the plenitude of legislative power resides in the States-General presided over by the King; that this august assembly has the right, not only to grant or to reject taxation, not only to draw up vain records of grievances, but also to examine the laws and how they can be reformed, and to make general regulations on legislation, on their own internal government, on trade, as well as on taxation. As the King has sole and undivided executive power, it belongs to him alone, after having concurred in giving the law the necessary approval, to decree it and to order its

> publication, to watch over its execution, and to pursue and punish those who break it.

In order that the States-General might secure for the nation the benefits that were expected, certain conditions, said Lavoisier, were essential for its assembly: the King's solemn promise that there would be no *lettres de cachet* (arbitrary warrants of imprisonment under the King's seal), no order for the imprisonment or exile or exclusion of any of its members; liberty of the press in all political matters, with the precaution that writers must sign their articles so that they could be prosecuted under the ordinary laws for any libel that they might commit; and, lastly, a pledge to convoke the States-General regularly, according to their own free decision, every three or five years. In one place in the memoir there is a vague, but implied, suggestion that the new organization proposed ought to be called the National Assembly rather than the States-General, since its origins were the Provincial Assemblies. "We shall, therefore, not take as our guide," he wrote, "what our fathers did, for they were wrong; we shall not travel along the road of ancient abuses; the time of enlightenment has come and we must now speak the language of reason and claim those human rights that are inalienable."

During these last months of 1788 Louis David (1748-1825) painted the portrait of Lavoisier and Mme. Lavoisier reproduced here as frontispiece. It was until recently regarded as the only authentic portrait of the great chemist. David's receipt, dated December 16, 1788, for the sum of 7,000 livres is extant. The portrait shows Lavoisier seated at a writing table and turning to Madame Lavoisier who leans on his shoulder with her left arm. On the table are some pieces of chemical apparatus for experiments with gases; below there is a large round glass flask with brass endpiece and tap, resting on a plaited straw ring, and behind this is a large silver hydrometer. From the description on a passport issued to Lavoisier in 1792 during the Revolution to enable him to proceed from Paris to Blois, it appears that he was "five feet and four inches [French measure, which amounts to five feet and eight inches in English measure] in height, with brown hair and eyebrows, brown eyes, long nose, small mouth, round chin, ordinary forehead, and thin face." *

During much of this year of 1788 Lavoisier was occupied with the preparation of his *Traité* for publication. After a failure of the harvest, a severe winter set in and distress was everywhere. Events were, however, taking shape. A royal decree of January 24, 1789, gave double representation to the Third Estate and declared that every bailiwick should elect a number of representatives proportional to its population and the taxes that it paid, but said nothing about whether voting in the States-General would be by orders or by heads. The Government had not adopted Lavoisier's proposals.

Lavoisier left Paris late in February, 1789, to take part in the election of the representatives of his order for the nobility of Blois, and in the preparation of the *cahier,* or writ of grievances, and the instructions to be given to such representatives as they would elect to the States-General. He was appointed secretary and directed to draft the *cahier* for consideration.

In it he included the reforms that he had already advocated in the Committee of Agriculture and the Provincial Assembly of Orléanais and in the memoir on the

* Denis Duveen, *Isis,* XLII (1951), 233.

States-General. He also made recommendations for the maintenance of the liberty of the individual, "the first and most sacred of human rights," by the abolition of all those illegal processes, not recognized by the laws of France, by which a subject might be arrested, imprisoned, or exiled, without regular trial before the ordinary courts of justice. In the case of urgent necessity involving national security, he provided that the executive could be given greater powers, and that, if the States-General should be of the opinion that provisional arrest was sometimes necessary, any subject so arrested should be brought before the justices within twenty-four hours and tried without delay according to the ordinary laws of the realm. To insure freedom of the press, he proposed that all publications carry names of authors and printers as a protection to the subject against libel or other injury. Taxation should be imposed on all citizens throughout France without exemption, he suggested, according to their net incomes. The reform of legal administration should be effected by simplification and improvement in procedure and by the establishment in the country parishes of advisory councils to help litigants to settle their differences without resort to the courts. He advocated the appointment, without further delay, of a council of well-informed persons, including some who had studied English criminal jurisprudence, to study the reform of the civil and criminal codes. Control of and strict accounting for expenditure in all Government departments should be maintained, with suppression of all useless offices and sinecures, and the submission of detailed annual accounts of expenditure; the Minister of Finance and the Ministers of other departments should be accountable to the States-General for all moneys received by them. He further recommended greater stipends for the curés, most of whom were in a condition bordering on poverty; uniform systems of weights and measures throughout France; the appointment of a council to devise a national system of education; and the establishment of a national constitution by the States-General before they dispersed, with a law that they were to meet at frequent intervals of their own will, at times and places entirely within their determination, without the necessity of any concurrence by the executive power.

The *cahier,* as drafted by Lavoisier, was adopted on March 28, 1789, by the assembled nobles, who then elected the Vicomte de Beauharnais (whose widow became Napoleon's Empress) and the Chevalier de Phelines as their representatives to the States-General, with Lavoisier as deputy-representative in case death, ill-health, or other cause, might prevent the attendance of one or other of their appointed representatives. It is interesting to note that the *cahier* was published in the same year in a brochure of fifty-three pages.

Lavoisier regularly attended the meetings of the '89 Club, founded in Paris early in that year with the general object of furthering the cause of a free constitution. Its members included Bailly, Monge, and Vandermonde, from the Academy; Brissot, editor of the *Moniteur;* the Duke de Rochefoucauld, one of the first of the nobles to join with the Third Estate; the Abbé Sieyès, Lavoisier's colleague in the Provincial Assembly of Orléanais and author of the pamphlet which appeared early in 1789 and which began with the words, "What is the Third Estate? Everything. What has it been hitherto in the political order? Nothing. What does it desire to be? Something"; Dupont de Nemours, who shared his interests in economics; the Comte de Mirabeau, and many others prominent in French life. All contributed to the events of 1789.

There were over four hundred members and the Club in its early days was an important political society.

The States-General assembled on May 5 at Versailles with no ruling as to whether they were to sit as three orders or in one assembly, to vote by orders or by heads. The Clergy had a representation numbering 308, the Nobles 285, and the Third Estate 621. Necker was now Minister again, Loménie de Brienne having been dismissed on August 25, 1788, from a Treasury that was finally insolvent. The first dispute arose over verification of powers, the procedure by which every individual representative's right to sit in the assembly was verified. The Third Estate wished to verify powers in common with the Clergy and the Nobles, but these orders proceeded with separate verification, whereupon the Third Estate took no action except to send repeated invitations to the other orders for verification in common and refused to regard itself as constituted until this had been done.

After weeks of proposals and conferences and deputations, during which the Clergy were almost on the point of accepting verification in common, the Third Estate with some representatives of the Clergy verified their powers and constituted themselves the national legislature by adopting the title of National Assembly on June 17 and assuming full sovereign powers under the presidency of Bailly. On the twentieth, three days later, the members found their place of meeting closed in preparation for a Royal Session to be held on the twenty-second. They adjourned to a tennis court below the château and took the Oath of the Tennis Court not to disperse until they had devised a constitution for France. Meanwhile, some of the Nobles joined the Assembly.

At the Royal Session, postponed until the twenty-third,

the King declared himself in favor of separation of the orders and announced that he would, if necessary, govern by himself, without their help, in the interests of the happiness of his people. Necker resigned and many of the Nobles and more of the Clergy joined the Assembly. The King realized that he had lost and recalled Necker. But the Court now thought of force as a remedy; regiments, mostly of foreign troops in French pay, were concentrated on Paris under the command of the veteran Marshal de Broglie. Intimidation and "a whiff of grapeshot," it was thought, might solve the problem for the Court party. The Assembly asked for the withdrawal of these forces, but the King announced that the session of the Assembly would be transferred to some provincial town and dismissed Necker with the order to leave Versailles.

In Paris, tempers rose; there was a scarcity of work and of bread, but there was freedom of speech and freedom of the press. When the news of Necker's dismissal arrived on July 12, insurrection broke loose. De Broglie's troops were of uncertain temper, their officers disliked the responsibility thrust upon them, and Paris was given over to the rioters. The weak municipal authorities, nominees of the Crown, released their duties to the electors of Paris, some three hundred in number, who had been appointed to choose the city's representatives on the States-General, and who had remained in association in view of the disturbed state of affairs. They assembled at the Hôtel de Ville as the Commune of Paris. A civic guard was formed, but meanwhile the rioters, joined by deserters from the disaffected regiments and armed with weapons seized from the Invalides, attacked the Bastille on July 14; the ancient fortress and political prison of Paris fell after some hours of resistance. It was the first effective act in the Revolution and July 14, 1789, marks the end of absolute monarchy in France.

With Paris given over to anarchy, the King's position was hopeless. The next day he informed the Assembly that the troops would be withdrawn and he recalled Necker. Many nobles, hostile to the Assembly, left France. Emigration and rallying of forces hostile to the Assembly outside the frontiers of France had now begun. The three hundred electors were recognized by the Assembly as the effective municipal authority in Paris. Bailly was appointed Mayor and the Marquis de Lafayette, who was devoted to the cause of the Assembly, was appointed to command the civic guard, which was renamed the National Guard. The King came to Paris from Versailles on July 17 and confirmed these changes. Provincial cities and towns followed the example of Paris, elected their own local authorities, and organized sections of the National Guard. The former administrative officials fled or were driven from their posts; the peasants rose and burned châteaux, while their owners hurried over the frontier; the courts ceased to function; taxes could not be collected; the ancient order had ended in chaos.

Lavoisier had returned to Paris from Blois in April, 1789, and resumed his duties in the Academy, the Tax-Farm, and the Gunpowder Commission; and he had found some time to continue his researches. As Commissioner for Gunpowder and resident in the Arsenal, he had not escaped the anxieties of these difficult days. On April 27, there had been the riot at Réveillon's paper warehouse in the Rue St.-Antoine, hard by the Arsenal. It had originated in a report that Réveillon had said that a workman could live very well on fifteen sous a day; and, despite the intervention of troops, it had gone on until the mob dispersed under imminent threat of grapeshot from artillery ordered to the scene next day. Three weeks after the fall of the Bastille, he had, as we have already seen, narrowly escaped with his life from an infuriated and misinformed mob deluded by rumor. In September he was elected to the Commune of Paris for the district of Culture-St.-Catherine. His colleagues, besides Bailly and Lafayette, included Condorcet, de Jussieu, and other Academicians; de Gineau, Professor in the Collège de France; Demachy, the chemist; members of other academies and learned societies; together with Brissot, the barrister Danton, the brewer Santerre from the Faubourg St.-Antoine, the Farmer-General Duvaucel, and others. When the Civic Guard, later the National Guard, was formed, Lavoisier was enrolled in the section for the Arsenal. Some records have survived of his service: he was ordered for guard duty on August 10, 1792, at the Arsenal magazines and on May 31, 1793, at the Place de la Révolution.

Lavoisier's first direct contact with the National Assembly took place on July 20, 1789, when as Governor of the Discount Bank (originally established by Turgot to provide credit for trade) he went with his colleagues to convey the respects of the shareholders to the Assembly. The Duke de Liancourt, President for the day, congratulated Lavoisier on the patriotism of a finance company that had not suspended payment during the troubles that had recently disturbed Paris, and this was recorded in the proceedings. (It was well known that members of the company had on occasion used their private fortunes to maintain payment and that, in times of difficulty, Ministers had often made use of the Bank's resources to cover Government deficits.) At a meeting of the shareholders on November 17, Lavoisier presented a financial statement showing that the nation owed the company 60 million livres, to be assigned against the compulsory patriotic contribution decreed by the Assembly, namely, a quarter of every citizen's income for one year only. When Necker's plan for converting the Discount Bank into a national bank was discussed, a member of the Assembly asked for Lavoisier's statement to be clarified, and for the issue of his report to be prohibited if the suggestion about assignment had not been deleted, but the matter did not appear to raise any ill consequences. Lavoisier himself was committed to give 30,000 livres to the patriotic contribution; he paid 14,000 in November, 1789. In undertaking to pay the remainder during the course of the year 1790, he pointed out that the sum of 30,000 exceeded the proportion fixed by the decree of the National Assembly.

On August 4 the Assembly, now well informed of the spreading revolt in the provinces, in one night's session hastily resolved on a whole series of drastic and sweeping reforms, including, among many others, equality of taxation and the abolition of tithes, feudal dues, and feudal courts, and all privileges of one class or town or province over another. Their excitement and enthusiasm carried them too quickly along the road of reform; and they ignored the warning of Sieyès that the abolition of the tithe was a relief to every landed proprietor in France.

But the chief subject of discussion was the drafting of a constitution; and on this the Assembly were divided. On the extreme Right a small party wished to keep the old ways with periodic States-General, and on the extreme Left an even smaller group held republican views. The Right Center, or *Monarchiens,* wished to have a constitution on the English model, with the monarch having a wide executive power and the parliament controlling taxation and legislation. The Left, later known as the *Feuillants* from the meeting place, the Convent of Les Feuillants, of the political club to which they belonged, wanted much the same thing, but not on the English model and with less power in the hands of the monarch. Discussion proved abortive and the ideas of the Left gained ground.

In October, rumors spread in Paris that a military plot was being hatched at Versailles to get the King away to Metz, where he might be a rallying-point for political *émigrés* and others hostile to the Assembly. A shortage of bread in Paris led to a demonstration by a crowd of women at the Hôtel de Ville, which ended in their long march of eleven miles to Versailles, in far greater numbers than had appeared at first in front of the Hôtel de Ville. Through the night Lafayette followed with his National Guard and serious disorder was prevented. The women wanted bread; but Lafayette may well have realized that this was an opportunity to prevent the departure of the King. During the demonstrations at Versailles, in which bread was promised, the crowd had observed, and prevented, several attempts to get the royal carriages on the highroad, but not in the direction of Paris. The cry was raised that the King must come back to Paris with his loyal subjects to insure his protection from the plotters at the Court. The Royal Family was thus forced to accompany the procession back to Paris, where it took up residence in the Tuileries; bread and fifty cart-loads of corn came, too. "Courage, my friends!" exclaimed one of the marchers, "We shall not want bread now: we are bringing you the Baker, the Baker's Wife and the Baker's Boy." Until 1791, the King and Queen and the Dauphin lived in the Tuileries without molestation, while the Assembly took up their quarters in the Riding-School.

During this year of 1789, Lavoisier had seen the publication of his *Traité;* and he had been active in the foundation of a special journal for chemistry. The annual volumes of memoirs of the Academy were often some years in arrears and the monthly *Observations,* founded by Rozier in 1771, had now passed into the hands of Jean Claude de la Métherie (1743-1817), who was a partisan of the phlogiston theory and hostile to the new chemistry.

A young chemist, Pierre Auguste Adet (1763–1832), whom we have already mentioned as joint author with Hassenfratz of a new system of chemical symbols, had attempted since 1787 to obtain the necessary legal *privilège* to publish a new journal. This work would contain translations of memoirs on chemistry from the German *Annalen,* edited by Crell, together with others from English and French authors, and thus provide a medium of publication for memoirs on the new chemistry.

Despite Lavoisier's support, Adet failed to get anything more from the Keeper of the Seals than permission to publish merely a translation of Crell's *Annalen,* with the restriction that the work should appear only quarterly, either of which conditions would defeat the aims in view. Even a second attempt was unsuccessful, although Lavoisier stated that his colleagues in the Academy shared his opinions and although he gave very cogent reasons in support of the application. Adet was completely discouraged; he was talented; he had read three memoirs to the Academy; but he had no means. He thought of emigrating to the colonies. Lavoisier tried to get various appointments for him, but apparently without success and Adet lived in such straitened circumstances that Lavoisier had to come to his help more than once. On February 12, 1788, Adet borrowed 144 livres from Lavoisier and on August 9 an equal sum from Fourcroy; when he could not repay Fourcroy, Lavoisier paid the debt for him, and, in 1791, lent him a further sum of 600 livres. Adet's

ANNALES DE CHIMIE;

OU

RECUEIL DE MÉMOIRES

CONCERNANT LA CHIMIE ET LES ARTS QUI EN DÉPENDENT.

Par MM. DE MORVEAU, LAVOISIER, MONGE, BERTHOLLET, DE FOURCROY, le Baron DE DIETRICH, HASSENFRATZ & ADET.

TOME PREMIER.

A PARIS,

RUE ET HÔTEL SERPENTE.

Et se trouve à LONDRES,

Chez JOSEPH DE BOFFE, Libraire, Gerard-Street, N°. 7 Soho.

M. DCC. LXXXIX.

Sous le Privilege de l'Académie.

24. Title-page of the first volume of the *Annales de Chimie* (1789).

receipts for these loans were found in Lavoisier's papers. Later in his life, Adet became the first Minister of France to the United States.

In 1789, however, circumstances were more favorable for publication. Freedom of the press had come, in fact if not in law; and the first issue of the new journal, the *Annales de Chimie,* appeared in April of that year under the joint editorship of Lavoisier, de Morveau, Monge, Berthollet, de Dietrich, Hassenfratz, and Adet, who acted as secretary. The title page is reproduced in Fig. 24. Lavoisier was treasurer and effective editor. Thus, in the year of the Revolution, was born the oldest of our surviving chemical journals, owing its appearance, in effect, to the rising tide of popular feeling in France against all restrictions on liberty and, therefore, against the restriction on the freedom of the press. The *Annales* still appears, although publication ceased after Lavoisier's arrest in 1793 and was not resumed until 1797; meanwhile, two of its original editors, Lavoisier and Dietrich, had perished in the Terror. In its long life of 164 years it has played a great part in the advance of chemistry; its pages have contained many famous memoirs, and the list of its contributors has included some of the greatest names in chemistry.

There remains one other interesting document relating to this memorable year. On September 19, Lavoisier wrote to Joseph Black, sending to him by the hand of de Boulogne, who was leaving Paris for Edinburgh to complete his studies, a copy of the recently published *Traité.* "You will find in it," he wrote, "some of the ideas of which you sowed the first seed; if you will be good enough to give a little time to reading it, you will find in it the development of a new doctrine which I believe to be simpler than the phlogiston theory and more in agreement with the facts. It is, however, only with trepidation that I submit it to the foremost of my judges, whose approbation I seek above that of all others."

XXIV

TROUBLED YEARS (1790–93)

LAVOISIER, ALTHOUGH SATISFIED WITH WHAT HAD SO FAR been effected by the Revolution, began to be a little anxious about the turn of events early in 1790. He had now taken the oath at the bar of the National Assembly as deputy-representative of the nobility of Blois. Writing to Franklin on February 5, he said:

> After telling you about what is happening in chemistry, it would be well to give you news of our Revolution; we look upon it as over, and well and irrevocably completed; but there is still an aristocratic party, making some useless resistance and very weak. The democratic party is in the majority and is supported moreover by the educated, the philosophic, and the enlightened, members of the nation. Moderate-minded people, who have kept cool heads during the general excitement, think that events have carried us too far, that it is unfortunate that we have been compelled to arm the people and to put weapons into the hands of

every citizen, that it is inexpedient to put power into the hands of those who ought to obey, into the hands of the very people from whom it is to be feared that the setting up of a new constitution will meet with obstruction on account of what it has established. . . . We greatly regret your absence from France at this time; you would have been our guide and you would have marked out for us the limits beyond which we ought not to go.

Later in the year, on July 5, a letter to Black shows less clearly, but plainly enough, the same feeling of anxiety. "The state of public affairs in France during the last twelve months has temporarily retarded the progress of science and distracted scientists from the work that is most precious to them; but we must hope that tranquillity and prosperity will follow the troubles through which we have passed and which are inseparable from a great revolution."

Every day a new duty fell upon him, keeping him from his laboratory. There he had embarked on his last and one of his greatest researches, the chemistry of the human body, in which he advanced in a great stride towards modern physiology; we shall refer to this work in a later chapter. On all sides his assistance was sought: by the committee on money, concerning technicalities in the production of *assignats,* the paper money issued by decree of the National Assembly, and the methods of detection of spurious notes; by the committee on health; and by the committee on taxes for whom he prepared an extensive report on the import duties for Paris, which he presented in November. He was directed to examine, in collaboration with Quinquet, a process for the protection of gun barrels against rust; to make experiments on hospital hygiene for the committee on vagrancy; to be present with

Vandermonde, Cousin, and Carny, at the casting of guns. He served on the Committee of Weights and Measures and acted as its Treasurer; he protected the interests of the Academicians, whose salaries were irregularly paid; and he was called upon to settle a dispute between Hassenfratz and Séguin on the subject of researches on vegetation.

For July 14, 1790, when the Federation Fête, the "Feast of Pikes," was held in the Champ-de-Mars in Paris, he invited the representatives of the National Guard of Blois to stay at his house, received them at his table, and gave them a warm welcome, which earned the thanks of the inhabitants of Blois.

Ten days later, on July 24, however, he wrote again to Black introducing M. Terray, a French student proceeding to Edinburgh, and said: "The Revolution that is taking place in France must naturally make some of those attached to the former administration superfluous and it is possible that I may enjoy more freedom; the first use that I shall make of it will be to travel and especially to travel to England and to Edinburgh to see you there, to listen to you and to profit by your enlightenment and your counsels." There is a note of personal anxiety about the future not well concealed in his words.

He was active in the '89 Club. Its fortunes were high at this moment, but they were not rising. The Club's celebrations in the Palais-Royal and its distinguished guests brought cheering crowds, but membership slowly declined, while that of the Jacobin Club increased with affiliated societies throughout France. Founded to support a free constitution, the '89 Club eventually came to be regarded as having aristocratic tendencies and its members fell under suspicion of *incivisme,* of being opposed to the Revolution.

We know little of its last days. Lavoisier was secretary in January, 1791; only one volume of its journal was published; and it probably came to an end in the course of 1791. Later it became an offense to have been a member of the '89 and, when the Jacobins, early in 1794, decided on a "purifying scrutiny" of their list of members, the political "purge" of our modern age, they resolved to expel all who had been members of the "monarchical society of '89." Hassenfratz protested and, by boasting that he had been expelled from the '89, retained his membership in the Jacobins. The '89 Club could never have become popular. Its subscription for membership was five louis; it did not discuss current affairs, but problems of a general kind and what it called "political metaphysics"; it stated that it was "neither a sect nor a party, but a company of the friends of men or, as might be said, agents in the exchange of social truths." Its avowed objects cut it off from all influence on the mass of the nation, which was not interested in "political metaphysics"; its discussions were too academic and too remote at a time when the body politic of France was passing beyond the aid of metaphysicians.

While the '89 Club was still an important center of political thought and discussion, Lavoisier delivered to its members on August 29, 1790, a remarkable analysis (afterwards published as a memoir) on the *assignats* and on the liquidation of the national debt. The Assembly, attempting to govern with a bankrupt Treasury, was harassed with financial problems which proved a constant strain on their ingenuity. On November 2, 1789, they expropriated the Church lands, to which they later added other lands belonging to the Crown and to the *émigrés*. By decree of December 19, 1789, they issued interest-bearing *assignats* to the amount of 400 million livres on the security of the expropriated lands pending their sale. The *assignats* were redeemable in five years and were not legal tender except for the purchase of expropriated land. But sales were slow and revenue was declining; and there were debts to discharge. Finally more *assignats* were issued, but as legal tender and at lower rates of interest. The burden of debt harassed the new governors of France. Lavoisier, as economist and financier, had much counsel to give on these matters; and his address to the '89 Club was a superb presentation of the problem.

The only resource available for discharge of the State debt, it was generally agreed, he said, was the sale of national property, the expropriated lands; and this, it was likewise generally agreed, could be done in one or other of two ways. First, all titles of credit against the State could be regarded as exchangeable for such lands; and second, *assignats* could be issued as legal tender and equal in amount to the State debt and withdrawn successively from circulation and destroyed as this national property was sold. All data must, however, be considered and a real, not a theoretical, solution was required, applicable in all respects to prevailing circumstances.

Since every plan turned on the sale of the expropriated lands, it was, said Lavoisier, primarily essential to estimate their present value; varying figures had been quoted. His own estimate for the Church lands was 4 milliards of livres before expropriation but, since the abolition of the tithe and of tolls and the reservation, in the interests of a maritime nation, of the forests, so many of which were on Church lands, the estimate of their capital value must be reduced by two-thirds. This same adjustment must be made for the royal domains and other nationalized land; and the total capital value, he estimated, was not more than 1,800 million livres. Already, he said, the National Assembly had issued 400 million livres in *assignats* on the security of this property, while, during the next year, 1791, they would need for various purposes, to fill the gap of decreased revenue, to repay advances that would fall due, and to provide money for armaments ordered, 250 million livres by May 1. They would be faced during the last eight months of the year with a deficit in revenue which he estimated at 100 millions, giving a total of 750 millions. As the value of the land to be sold was, according to his estimate, 1,800 millions, the future available national assets, therefore, amounted to 1,050 million livres.

"You will be dismayed, gentlemen," he said, "to see that capital, valued at four milliards when the nation took possession, has been reduced in so short a time to one milliard; and perhaps you will regret that momentary enthusiasm led the National Assembly to renounce the tithe, the redemption of which would have contributed so effectively to national recovery and the extinction of the national debt."

Prudence would therefore counsel an issue of no more than two milliards of *assignats,* or preferably only one milliard; for, if future events were to prove that even a mere fraction of the *assignats* was counterfeited, or if such a thing could be as much as suspected, it would be the end of all credit.

Turning to the debt itself, Lavoisier stated that it had been calculated as 1,900 million livres. But in this calculation no discrimination had been made between the different kinds of debt, long-term and short-term loans not being separated but taken all in one total. By an analysis of the different debts and the exclusion from consideration of those payable at later dates in less difficult and less embarrassing conditions, he contended that the debt to be considered in the present problem amounted to 1,200 million livres. To issue 2 milliards of *assignats* would reduce the value of money by a half, all other things being equal, said Lavoisier, founding his argument on the economic teachings of Hume, but they were not equal, as he would presently show. The argument of the supporters of *assignats* was that, while the nation increased its currency by 2 milliard livres, land of an equivalent, or assumed equivalent, value was thrown on the market, and the one would balance the other without increase in prices. Lavoisier, however, pointed out that this argument assumed that the *assignat,* as soon as it was issued, would be cancelled by the acquisition of its equivalent in land, which was not possible, since buyers would take time to visit and inspect any property that they proposed to buy and to consider its suitability, and since some would wish to pay gradually, as they were permitted to do by decree of the National Assembly. Allowing for this delay, he concluded that the currency would not be effectively doubled and that the consequent rise in prices would perhaps be only a third or a quarter, according to the progress of the sales of land.

Without exaggeration, one might therefore say that prices would rise by about one quarter, which would be comparable to the imposition of a tax of 25 per cent. And how would the French manufacturer then compete with the foreigner? French trade would be utterly ruined. If the reply to this was that a rate of exchange of currency would presently be restored with other nations, he would agree, provided foreign imports were paid for in effective currency, which would have to be so, since the foreigner would not accept payment in paper money that had only a representative value. But France would then soon be short of effective currency and there would be available only paper money extinguishable by the sale of national land; and France would have neither effective currency nor enough paper money.

Who could assess the disasters that would follow this double crisis? A half-century might enable France to surmount it by the sacrifice of the happiness and prosperity of two whole generations, but was it for this that France had nominated her representatives? So large an issue of *assignats* was, therefore, concluded Lavoisier, useless. The result desired could, however, be achieved in another way without gambling so hazardously with public fortune and individual well-being.

Lavoisier pointed out that the creditors of the State were not easily identifiable. They were lenders to the State who were often borrowers from other lenders in an infinite ramification extending through all grades of society. To free the State from its creditors, according to the method proposed, could not be done with justice without allowing these creditors to be able to free themselves in turn from their creditors; and it was, in fact, a very difficult and complex problem. These considerations, he observed, provided a further good reason for distinguishing between the different kinds of State debt. His plan was, therefore, to convert all titles of credit against the State into *quittances de finance,* or bonds, not payable on demand but bearing an annually decreasing rate of interest, which would be used in the purchase of land. Thus the State's debts would be extinguished in free and voluntary exchange for land. All bonds not used in such purchases in the first year, Lavoisier proposed, would be repaid in four equal payments over a period of four years, at the rate of two to three million livres annually, as the money for these repayments became available by progressive sales of land.

He had already shown that the needs of the Government in the next year would demand a new issue of *assignats* to the value of 350 million livres. Further, it would be necessary to put another issue of the value of 500 millions into circulation during the last months of 1790 and in the course of 1791. An issue of about 900 millions by the end of 1791 could be expected. Since this was already an amount greater than the circulation could carry, bearers would doubtless not delay in exchanging them for land. The *assignats* paid back into the bank during 1791 would provide a considerable sum for the repayments due in the course of the following year, and so on. There were thus three great financial operations to be carried through together in 1791: first, the successive issue of *assignats* in proportion to public needs to a value of 500 million livres, which, added to the 400 millions already issued, would make a total of 900 millions; second, conversion of 1,200 millions of the outstanding debt into *quittances de finance,* repayable in *assignats* from 1792 to 1795 at the rate of 300 millions annually, the total, however, being less by the payments made during the first year, 1791; and third, the sale of the expropriated lands, which would be spread over the year 1791 and would provide the money for repayments during 1792.

Lavoisier's plan differed from those of others in that he proposed (1) to spread repayment over four years, whereas others had proposed to complete the operation in one year, (2) to keep the repayment of the national debt down to the lowest possible figure, and (3) to allow those who held titles of credit against the State to be free either to exchange their credits for land or to retain them until the date on which they would be due for repayment with the accumulated interest. He recognized that his plan had much in common with the proposals of others, but he emphasized its special feature, namely, the restriction of the issue of *assignats,* in progressive and slow stages according to conditions, to the least possible total, estimated at 800 to 1,000 millions. He completed his memoir by presenting a detailed draft of the decree that would be necessary.

These careful and cautious proposals, while they might be well adapted to solve the problem, did not, however, recommend themselves to the National Assembly. Moreover, the revenue declined further and reforms that were decreed involved more heavy expenditure; and again and again the Assembly issued *assignats,* the total eventually reaching 1,800 milliards with consequent depreciation of paper money. Necker, the Minister of Finance, himself became discredited, resigned, and left France finally on September 3, 1790.

But while this well-devised scheme was being presented by Lavoisier, mutiny broke out among the troops at Nancy. The Army had become increasingly disaffected since the Federation Fête in July, when representatives from the National Guard had come to Paris from all parts of France to attend the celebrations in the Champ-de-Mars and had thereafter frequented the political clubs. Although the mutiny was suppressed, with heavy casualties, it had shown that there was much dissatisfaction among the soldiers. They were badly fed, poorly clothed, and infrequently paid. Regimental committees were formed to give voice to their grievances, but their officers, mostly nobles, were indifferent to the interests and comfort of the rank and file, and there was no promotion for non-commissioned officers. The Federation Fête had spread the spirit of revolution throughout France. Officers began to leave to join the *émigrés* across the frontier; men deserted their regiments. Discord among the civilian population, as well as discontent in the Army, increased. Pamphlets and journals multiplied and vied with one another in denouncing the unremedied condition of the people. Prominent among these journals was the *Ami du Peuple,* edited by Marat, the self-styled "People's Friend."

Jean Paul Marat was a fanatic; he was sincere; he was violent and bloodthirsty; he was enthusiastically and lyrically devoted to the Revolution; and he was probably insane. If the Revolution were not succeeding, if the great promises of the future were not being fulfilled, there must, it seemed to Marat, be treachery somewhere and it must be uncovered from its secret places. To picture this man as a mere hunter of scapegoats is to misunderstand him; he really believed what he said. In his earlier days, he had been ambitious for scientific reputation. In 1780 he had published a book on his physical researches on fire, *Recherches physiques sur le Feu;* it was a work devoid of all merit and it claimed that its author had succeeded in rendering visible the element of fire. The *Journal de Paris* incorrectly reported that the book had received the approval of the Academy of Sciences, but Lavoisier, on reading this statement, made a brief public announcement in a few scornful words on June 10, 1780, that the Academy had not given its approval to the work and that Le Roy had been instructed to reply to the statement. The *Observations* for July, 1780, remarked that the book contained some singular ideas about the nature of fire, but that it would be expedient for scientists to repeat the experiments described in it in order to verify them and to ascertain of what use they were. Thus, it appeared to Marat, opposing forces had prevented him from achieving scientific distinction; these same forces—even some of the very same men, Lavoisier and other Academicians—were preventing the realization of the "Rights of Man."

The *Ami du Peuple* for January 27, 1791, appeared,

therefore, with a calumnious and vulgar denunciation of Lavoisier by Marat. Before we quote it, it is perhaps worth recalling that Marat had seen nothing but treason, delusion, and debauchery, everywhere, and that his solution was the erection of eight hundred gibbets in rows and the hanging of Mirabeau on the first of them. Marat wrote: "I denounce to you the Coryphæus—the leader of the chorus—of charlatans, Sieur Lavoisier, son of a land-grabber, apprentice-chemist, pupil of the Genevan stock-jobber [Necker], Farmer-General, Commissioner for Gunpowder and Saltpeter, Governor of the Discount Bank, Secretary to the King, Member of the Academy of Sciences. Would you believe that this little gentleman, who enjoys an income of 40,000 livres and whose only claim to public recognition is that he put Paris in prison by cutting off the fresh air with a wall that cost the poor 33 millions and that he moved gunpowder from the Arsenal into the Bastille on the night of July 12 and 13, is engaged in a devilish intrigue to get himself elected as administrator of the Department of Paris. . . . Would to Heaven that he had been strung to the lamp-post on August 6. Then the electors of La Culture would not need to blush for having nominated him." There was more to follow in a similarly outrageous pamphlet, entitled *Modern Charlatans, or Letters on Academic Charlatanism, published by M. Marat, the Friend of the People,* in which he attacked the most distinguished of the Academicians, Laplace, together with Monge, Cassini, Beaumé, and Fourcroy, and accused them of having spent in debaucheries the 12,000 livres given by the Government for the experiments on aeronautics!

But Lavoisier was the object of his greatest violence: "Lavoisier," he wrote, "the putative father of all the discoveries that are talked about; as he has no ideas of his own, he seizes those of others; but scarcely ever knowing how to appreciate them, he abandons them as lightly as he took them up, and changes his views as he changes his shoes. In the space of six months, I have seen him in turn lay hold of the new doctrines of elementary fire, of igneous fluid, of latent heat. In a shorter time, I have seen him infatuated with pure phlogiston and unsparingly pursuing it. Proud of his achievements, he sleeps on his laurels, while his toadies laud him to the skies."

The Gunpowder Commission, too, was criticized in the political clubs and the newspapers, and a demand arose for its abolition. Lavoisier, on behalf of the Commission, wrote to the Ministers of War and Marine, giving an account of the situation, of the work of the Commission, and of the available stores of powder. To the public the four Commissioners addressed a pamphlet, drawn up by Lavoisier, summarizing their work and their success since the establishment of the Commission by Turgot in 1775, and defending themselves against the calumnious and anonymous attacks in the papers and the gossip and rumor in the clubs. Facts, easily verifiable, were presented; and the Commissioners excused themselves for the personal details that their pamphlet contained on the ground that their patriotism and their honor had been attacked and that they must therefore defend themselves. With regard to the demands for the abolition of the Commission and for the free manufacture and sale of gunpowder, they pointed out that the production, storing, and sale, of gunpowder had been restricted to the sovereign power, no matter what form that power might take, from the time when the use of firearms had become general in Europe. Such restriction was essential to the safety of nations. It was not a fiscal law, but a law for public and individual security, and one which had been made against the great and powerful, who alone could corner the supplies of gunpowder, and from whom the people might fear oppression.

They added: "Legislative power would be nullified if it lacked the force to repress disobedience, or if disobedience could at will multiply the means of resistance. The free accumulation of arms by individuals would disturb the tranquillity of all; the accumulation of gunpowder would be even more dangerous and more disquieting, since it could produce appalling effects without awaiting the order or consent of the very people who had permitted it. Bread is no more necessary for the survival of individuals than gunpowder has become for the continuance and tranquillity of empires." Further, they pointed out that such a change as that advocated would bring about a return of the old conditions, when France depended on the foreigner for most of her gunpowder, and when England, sure of her supplies of saltpeter from India, could hold France to ransom. France was now in this matter of gunpowder and saltpeter the only self-supporting power in Europe, the only one not ultimately dependent on India; and, indeed, it might have been added, able to export gunpowder to her allies, as, for instance, to America during the War of Independence.

But the old rumors had been raised again and given new currency: the movement of powder into the Bastille on the night of July 12–13 at the instance of the Governor, who was, in fact, acting under superior orders, and the removal of powder from the Arsenal on August 6, 1789 (see p. 187), and so on. It was all very disquieting to Lavoisier.

In the midst of these attacks and anxieties, Lavoisier on March 15 presented to the National Assembly an abridgment of the work on which he had long been engaged, *The Territorial Wealth of the Realm of France*. He had been preparing this study since 1784, and had edited the abridgment for the Committee on Taxation appointed by the Assembly. The Assembly, favorably impressed with the book, ordered it to be printed; it appeared in 1791.

For economists who wished to establish a system of taxation based on the net product of the land, it was essential to have exact returns of products and to draw up an account of the agricultural wealth of France. Dupont de Nemours had attempted this in 1784 and had arrived at the figure of a total production of 2½ milliard livres for a population of 20 millions, an average of 125 livres per person. He was, however, not satisfied with his figures and thought that the total value of the annual products was nearer 3 milliards, giving an average of 150 livres. Voltaire in 1767, in his *Man of Forty Crowns,* had given a gross yield of 2 milliards 400 millions, with an average of 120 livres (or forty crowns, whence the title of his book). Lavoisier gathered his data for all the provinces of France through officials of the Tax-Farm and aimed at the compilation of a complete return for all that France produced. Data of this kind and calculations from them, he said, were the basis of all political economy; the science of economics had begun in metaphysical argument and discussion, and practical economics had not been developed, while statesmen continually lacked the necessary facts. The representatives of France would have found their task much lighter if this information had been collected by the preceding generation; now they had an opportunity of founding an institution before all other nations for the compiling of such statistics from year to year —on agriculture, on commerce, and on population, an annual epitome of the national wealth in men, in produce, in industry, in accumulated capital.

Developing his theme, Lavoisier explained that every product would have its own special chapter. Agriculture would be considered as if it were the domain of a single individual charged with the receipt of all its products and accountable for what was done with them. For example, there would be a chapter for corn, for the entire national harvest, amounting to fourteen milliard livres, and the whole of this return would be rearranged in an account of expenses in kind under different heads: the amounts respectively returned to the farmers for seed, set aside for the farmers' subsistence through the year, given to the harvesters for their labor and to the threshers for their work, taken by the tax-collectors, paid as rent to the landlords, and so on. Every product would have a chapter on these lines with figures in kind; and there would be a corresponding chapter in terms of the equivalent values in money. This had been his plan at the beginning, but he had completed only a small part of it. "Twenty times," he said, "I have returned to this work and twenty times I have left off and, although I knew how important it was, although I wished to publish the results in time for the Committee on Taxation to make use of them in deciding on the taxes, continually deflected by tasks of another sort, many of which are not unknown to the National Assembly, I have been altogether unable to complete it."

According to the figures presented by Lavoisier, the value of the total annual yield of the soil amounted to 2 milliards 750 million livres for a population of 25 millions, giving an average of 110 livres, a figure much below that of Dupont de Nemours and slightly below that of Voltaire. The net yield, the value subject to taxation, Lavoisier estimated, amounted to 1,200 millions with corn at the price of 2 sous per livre. But, at this time, the price was below 2 sous; and the yield would therefore not exceed 1,000 millions. The results showed, therefore, on taking a mean between these figures, that the budgetary proposals of the Assembly for a tax of one-sixth would not provide, even with the estimated addition of 30 millions from the towns, a revenue of 240 millions, but of 210 millions, leaving a deficit of 30 millions.

The extensive tables presented by Lavoisier included details of the population by age, sex, and occupation, the numbers of farm animals, the consumption of corn, oats, rye, meat, and wine, the acreage of arable land. National statistics were available for the first time, and were based on reliable returns, which gave a knowledge of the national wealth and of what tax could be imposed. The scientific and detailed presentation of the data and the conclusions drawn from them made a profound impression on the Assembly. This research entitles Lavoisier to rank as an economist of distinction.

The *Moniteur* of May 26 said that the work "comprised in small compass the whole basis of political economy", and added that "to increase knowledge on a subject so intimately linked with public prosperity is to render very useful service to one's country." One of Lavoisier's comments on his study of the population statistics proved particularly impressive in the public mind in showing the small importance of the nobility in comparison with the *Tiers État*. "It is seen [from the tables]," he wrote, "that the former nobles, including those raised to that rank, constitute only one three-hundredth of the population of the realm and that their number, men, women, and children included, is merely 83,000, of whom only 18,323 are fit to bear arms. It is seen also that the other classes of society, those whom it was usual to combine under the name of *Tiers État,* can muster 5,500,000 men fit to carry arms."

Public clamor about the Tax-Farm began to increase. From the time when the States-General met in 1789, subordinate employees had voiced their complaints and the excise clerks of Paris had sent petition after petition to the National Assembly, demanding an account of their pension fund and complaining about the way in which it was administered. When the Farmers-General, no longer able to ignore these attacks, sent to the Controller-General in November, 1789, a statement on the pension fund, the clerks replied with a pamphlet couched in terms of extreme violence and hatred: "Where can one find more cruel masters! . . . Ah! if we could examine the accounts supplied to the Government by the Farm, what mysteries unknown to the State would be unfolded! . . . Tremble, you who have sucked the blood of the unfortunate and deceived the most beneficent of kings!"

Other employees published pamphlets defending the Farmers and asserting that the instigator of the excise clerks' petition was an official of the Farm who had been dismissed for theft. But it was all useless, and further attacks of even greater violence followed, together with the new suggestion that the fortunes of the Farmers-General should be confiscated by the State. "Our opponents dread the scrutiny of the documents, the publication of which they persist in refusing to us, documents all the more interesting in that they would lead to the return of millions to the national exchequer." The *Patriote français* called for the suppression of the Farm, "that body whose long-desired annihilation is not far distant." The National Assembly, after abolishing the *gabelle,* gave way to the agitation, and on March 20, 1791, annulled the Tax-Farm.

Père Duchesne pointed the Farmers-General out to his readers as objects for the vengeance of the people, warned all to keep an eye on them and to make them disgorge what they had gained by robbery and extortion.

The decree of suppression was retrospective and applied from July 1, 1789, so that all acts of the Farm since that date were to be regarded as done in the name of the nation. This greatly complicated the completion of the accounts and the liquidation of the Farm's affairs, duties entrusted by a clause in the decree to six of the former Farmers-General and three assistants. The six chosen for this task did not include Lavoisier; thus on March 20, 1791, he ceased to play a part in the organization to which he had belonged for twenty-three years.

"Bad days had come upon him," wrote Grimaux; "When he lost his great place in the financial company in which he had played his part for more than twenty years, his fortune and presently his liberty in danger, he was grieved at not finding with the Ministers the support that he had a right to expect from them." The Committee on Taxation proposed his appointment by the King as one of the administrators of customs, but he was deeply disappointed to find that his name had been removed from the list submitted for the King's approval; and his own application for a place in the administration of the Paris excise was not successful. It is possible, however, that the Ministers had other and more important duties in mind for him. In the early days of April, when the control of the Royal Treasury was taken over by the nation, the Minister of Finance nominated a commission of six, including Lavoisier, to be responsible for the exchequer. On Lavoisier's proposition, the Commissioners asked for the Royal Treasury to be renamed the National Treasury; the Assembly accordingly issued a decree authorizing the change.

Lavoisier was still a Commissioner for Gunpowder and he did not wish to accept two salaries. He wrote to the Minister and inserted the letter in the *Moniteur* of April 7: "I request that I may be allowed to carry out the new duties entrusted to me without salary. The emoluments that I enjoy as Commissioner for Gunpowder, for the very reason that they are modest, suit my way of life, my tastes, my needs, and at a time when so many good citizens are losing their positions, I could not under any circumstances agree to accept a double salary." Grimaux thought that Lavoisier, by inserting this letter in the *Moniteur,* hoped to disarm the enemies who had begun to attack him.

He was especially eager to retain his place as Commissioner for Gunpowder; but, when the Assembly reduced the Commissioners from four to three, he was chosen for dismissal on the grounds that he was in office at the National Treasury. He reminded Tarbé, Minister of Public Revenues, that he had accepted only provisional appointment in the Treasury and he asked that, until he could resume his work on the Gunpowder Commission, he might be allowed to continue to live at the Arsenal, where he had installed and equipped a laboratory at great expense. On October 1 he wrote to Tarbé: "I shall count on the verbal assurance that you have given me, in the King's name, that my right to a place as Commissioner for Gunpowder will not be set aside and that I shall have the opportunity of filling one as soon as a vacancy occurs. I shall count equally on your assurance about the residence that I occupy, where I have established myself at my own expense and where I have laid out much money in a laboratory, in collections concerned with the sciences that I pursue, and in instruments." Lavoisier's request was favorably received. The King authorized him to retain his accommodation in the Arsenal and he remained there for some time.

Political affairs were becoming ever more disturbed. On February 28 the castle-prison of Vincennes, a smaller Bastille, was attacked; Mirabeau died on April 2; the ceremonies in the Champ-de-Mars on July 17 ended in disorder and musket-fire. Elections for a new Assembly, now called the Legislative Assembly, produced a parliament that was mainly Left and, although not avowedly so, republican. The fires of discord were continually fanned by the pamphleteers; and the problem of finance bore heavily on all. Moreover, on the night of June 20, the King, accompanied by the Queen and the two royal children, departed unseen from the Tuileries in flight to the Army of the East at Metz. There he hoped to establish himself while awaiting either foreign help or reaction at home; but on the next evening the fugitives were arrested at Varennes and brought back to Paris. It was no longer possible to maintain the fiction that the Revolution had the support of the King.

The new Assembly, elected to office on October 1, numbered 745; as no member of the previous house might sit in the new one, it was composed of much youth and much inexperience. On the Right were about 160 and on the Left about 330; the remaining 250 mostly supported the Left. Membership included only a few nobles and a few clergy; the majority were from the middle class. There was much influence and control from outside, especially from members of the old Assembly who had foolishly passed the self-denying ordinance that excluded them from election to the new one.

But the harassing problem was the state of the national

finances. As one of the Commissioners of the National Treasury, Lavoisier had enhanced his reputation through the order that he had established in the accounts, by which the situation was known from day to day. At the end of November 1791, he was called to the Assembly to present a statement of receipts and expenditure. Lavoisier was also concerned with the situation for 1792 and 1793. He set forth his views in a work published under the title of *The State of Finances in France on January 1, 1792,* which did not bear his name, but was described as "by a deputy-representative in the National Assembly." The contents could not conceal his identity, however, since he referred to his authorship of the *Territorial Wealth of the Realm of France.*

In introducing his subject, he wrote: "At a time when all is exaggerated, good as well as bad, when everyone sees things through spectacles that either magnify or reduce them, that either set them at a distance or bring them near, when no one sees them either in their actual dimensions or in their proper place, I thought that it would be useful for someone to attempt to discuss the state of affairs objectively and to submit the finances of the State to strict arithmetical reckoning. . . . Let no reader expect to find here any interest other than that arising from the importance of the subject: this work will be as cold as reason." He proceeded to give details of current income and expenditure and estimates of these same items for 1792 and 1793; all was founded on figures as exact as could be obtained. Monthly returns of each were studied; the various kinds of debt payable on demand were classified—those payable to bearer, those due on stated days, others due after a definite period, those of an indefinite maturity, and those still to be evaluated. From this careful analysis, he estimated the debt payable on demand during the different years that he was considering and the probable revenue, together with the question of further issues of *assignats* and the sale of land not yet disposed of. A modern economist has described his discussion of the debt as superb.* But few of France's latest governors were inspired by the cold light of reason.

The year 1792 opened with even greater anxieties than those of 1791. The new Assembly had proved no more successful than their forerunners. Their views grew more openly republican, while the King vetoed many of their decrees and their hostility towards him increased. Some representatives wanted war to spread the doctrine and principles of the Revolution, others wanted peace because defeat might undo the achievements of the Revolution and victory restore the monarchy. The turn of events, so different from what he had hoped for, the lamentable state of the national finances, and the menace of war, were profoundly disturbing to Lavoisier.

To Nompère de Champagny, a relative of Mme. Lavoisier, he wrote on January 10: "What we feared about the receipt of revenue is proving only too true and payments to the public treasury of the taxes on land and property are very poor. On the other hand, we may be dragged into war by the over-hasty measures of the National Assembly, although it interests no one and no one has any wish for it. What is happening to us is, however, what has been observed in all popular governments; they recognize no limits at all. They ought to have waited for greater tranquillity in a philosophic constitution which declared that it wished to attack none." And six months earlier, his contemporary Priestley, as a result of a dinner held in Birmingham on July 14, 1791, to commemorate the Fall

* S. E. Harris, *The Assignats* (Harvard University Press, 1930), 30.

of the Bastille—a dinner at which Priestley was not present —had been fortunate to escape with his life from a drunken mob which destroyed his house, his library, and his laboratory. Lavoisier had joined with other Frenchmen in sending his sympathies to Priestley; and the Assembly had conferred honorary French citizenship on Priestley, Wilberforce, Bentham, and others, and had invited Priestley to take his seat among them as representative of the Department of the Orne. Priestley, accepting the first honor, declined the second in a letter written on January 13, 1792, on the grounds of his insufficient knowledge of the French language and of French affairs.

In this same month, Lavoisier was nominated a member of the Bureau of Consultation of Arts and Crafts, instituted in 1791 to advise the Government on useful inventions and to reward inventors; but, as he wrote to the "Society of Friends of the Constitution" at Blois, he was overwhelmed with work and began to feel the weight of the immense burden that had been laid upon him. He was short of time for even his experimental studies, and he wished now to abandon all public office and to limit his services to the nation to unpaid work. In February he resigned as Commissioner of the National Treasury, to the great regret of his colleagues, who expressed to the Minister their realization of what they would lose in Lavoisier's great capacity, good intentions, and untiring activity. Meanwhile a vacancy had occurred on the Gunpowder Commission and he was immediately recalled, but he no longer desired to re-occupy his old post. Officials nominated by the King had become suspect and the Commissioners for Gunpowder were more and more exposed to slanderous charges. Prudently he wished to abandon all paid offices where he might be a public figure, says Grimaux; but it is clear also that he was opposed to the unconstitutional practices of the Government. Indeed, he had scarcely resumed superintendence of the production of gunpowder when he submitted his resignation to the Minister, Clavière. He undertook, however, to continue his researches on the manufacture of gunpowder and the analysis of saltpeter. "The State," he wrote, "will have four Commissioners instead of three; I shall thus reconcile my duty and my principles and I shall serve my country as a free and independent man who wishes to hold no authority."

In March the King formed a new Ministry from the moderate Republicans, the Girondins, so called because most of their leaders came from the Gironde district. The trend of all affairs was more and more republican; and it began to be evident even in the Academy of Sciences, where hitherto men of science had described and discussed their work apart from the storms of politics. At the meeting of April 25, Fourcroy proposed that the Academy should follow the example set by the Society of Medicine and remove from its list of members all suspected of *incivisme,* and that the list should be read and the names of such members removed from it. For the Academy this was the first of a series of events that finally led to its suppression, the history of which we shall relate in a later chapter. We note here that the menace fell heavily upon Lavoisier because he bravely constituted himself the public defender of the Academy in its last days against the assaults of ignorant legislators bent on its destruction.

On April 20 France declared war on Austria—the Assembly because they saw in the Emperor of Austria the representative of feudal Europe, the enemy of the Revolution, and the protector of the *émigrés;* the King because victory might restore the fortunes of the monarchy. The opening engagements in the Netherlands were disastrous and the Ministry was dismissed on June 12. The armed populace of Paris invaded the Tuileries and the Assembly on June 20, but the King with great courage refused to make any promises to the rioters concerning the abandonment of his veto or the appointment of "Patriot Ministers." His behavior restored to him much of the public esteem that he had recently lost and led to a remarkable demonstration of loyalty. The hopes of the King, and more especially the Queen, were based, however, not on the successful working of a remodelled constitution, but on armed deliverance by the help and intervention of foreign powers friendly to their cause, which even at this late date was not wholly lost and might have been saved. But when the Girondins offered its salvation in return for a Ministry formed from among their party, the King refused, and thereupon the Left combined to make an end of monarchy.

In this turmoil Lavoisier wished to abandon all public offices and to resume his scientific work, in which he hoped to recover the peace and happiness of earlier years. He declined an invitation to become Minister of Public Revenues, a post to which the King wished to appoint him in appreciation and recognition of his knowledge of finance, his integrity, and his devotion to constitutional government. Writing to the King on June 15, he explained that he was convinced that the Legislative Assembly had gone beyond the limits prescribed by the constitution and that a Minister, faithful to that constitution, would therefore be in conflict with the Assembly and would end by being the victim of his principles and the source of new misfortunes to the nation. He begged to be allowed to serve his country in less distinguished posts, where he could render services, perhaps more useful and probably more lasting, especially in awakening the nation to a sense of its duties towards education, a subject upon which he was now closely engaged.

On the night of August 9–10, a new Commune of Paris was set up which usurped the former municipality; and on the following day an armed mob invaded the Tuileries. The King and Queen took refuge in the Assembly,

25. Lavoisier's house at 243 Boulevard de la Madeleine. The house stood opposite the Madeleine and has been demolished.

where they listened to a discussion of their fate and that of the monarchy. The final decision was that the King should be "suspended" from office and a new assembly, a National Convention, should be elected to make a new constitution. In the meantime, the Commune of Paris really ruled France. The Royal Family was placed in their custody and imprisoned in the Temple.

Lavoisier, having unofficially continued his work on gunpowder for some months, had informed the Minister that he would resign from his duties on August 15. On that day he left the Arsenal and moved to No. 243 Boulevard de la Madeleine (Fig. 25). The house was, however, not ready for immediate occupation and for some weeks Lavoisier and his wife resided at 18 Rue Mirabeau. During the autumn and winter of 1792–93, they leased a furnished apartment in a house in the Rue du Calvaire, St.-Cloud, presumably as a refuge in case of trouble in Paris.* The great laboratory, which had seen the birth of modern chemistry, was dismantled; the vast collection of instruments and apparatus was packed and removed, never to be used again.

Lavoisier's fears had proved correct. Three days after his departure, the commissaries of the local section entered the Arsenal, sealed all papers, and resolved to imprison the three Commissioners. The eldest, Le Faucheux, took his life in despair; five days later the Assembly ordered the raising of the seals and the release of the prisoners.

In September the elections were due and the Commune of Paris wished to terrorize the populace by the slaughter of those who had been imprisoned under suspicion of royalist plotting and the like. When the Austrians crossed the French frontier and captured French towns, the Commune called for the enrolment of all able-bodied citizens in the Champ-de-Mars on September 2. While 60,000 volunteers were being enrolled, appointed representatives of the Commune visited the prisons of Paris and in four days flung some fourteen hundred victims to be massacred by the mob after swift trial for a minute or so to decide whether the prisoner was a royalist plotter or not. The Commune then invited other cities to follow their example.

The election yielded an assembly, the National Con-

---

* Denis Duveen, *Isis,* XLII (1951), 233-34.

vention, which was formed, in part, of many who had sat in one or other of the two previous assemblies. One element of this body was composed of the fanatical Jacobins, who numbered about 100 and included Robespierre, Danton, Desmoulins, and Marat. The Jacobins had separated from the republicans of the Right, the Girondins, and now desired their overthrow. Another of the aims of this group was the execution of the King. They came to be known as the Mountain, because they sat on a raised part of the house. The Girondins numbered about 180; the other members, about 500, voted now one way, now another, and occupied seats on the floor of the house, whence they came to be known as the Plain, or derogatively as the Marsh. It was a house republican in its aims. At its first regular meeting on September 21, one day after its official opening, it abolished the monarchy. One day later France formally became a republic.

Meanwhile history was being made elsewhere and in a somewhat unexpected manner. In the engagement known as the "Cannonade of Valmy", the *sans-culottes,* in their first action as the armies of revolutionary France under the leadership of Kellermann, repulsed the Prussians commanded by the Duke of Brunswick at Valmy in the Argonne on September 20. Accounts of the progress of this action are incomplete and unreliable, but the halting of the Prussian charge short of its objective is not unreasonably believed by some to have been caused by the accuracy and intensity of the French fire. This increased efficiency may well be regarded as one result of the improvements, effected by Lavoisier, in the quality and range of French gunpowder during the preceding seventeen years. Although the real cause may never be known, the ragged soldiers of France brought the Prussians to a standstill and saw them withdraw before their fire. The whole army, generals and soldiers alike, became inspired with a new confidence; and on November 6 Dumouriez defeated the Austrians at Jemappes.

While Valmy may have been a victory lightly won, it established the Republic as a military power and its effects were not to be undone until twenty-three years had passed to that long day of June 18, 1815, at Waterloo. To these early successes and to the far greater victories of Napoleon's armies, Lavoisier's long scientific and administrative services as Commissioner for Gunpowder must have contributed much. Indeed, he was aware of the military position, when he reported to the Minister in December: "The powder-mills and magazines of the realm are well supplied; the department is operating at its greatest capacity; and the precautions that I have persuaded you to take enable France to maintain the most formidable war."

Lavoisier had left Paris for Fréchines when the Academy went into recess for the summer, returning for the opening meeting of the new session on November 14. It was his last visit to Fréchines.

Early in December the Convention formally constituted itself a court to try the King. He was arraigned on December 3; eight days later the indictment was read to him before the Convention. The trial began on December 26; on January 15 Louis was declared "guilty of conspiracy against public liberty and of attacks upon the general security of the State." In a long session of January 16-17 the death-penalty was decided on, and he was accordingly condemned on January 20 and guillotined on January 21 in the Place de la Révolution, formerly Place Louis XV and now Place de la Concorde.

Such was the grim opening of the year 1793. The progress and propaganda of the Revolution and the execution of the King now brought England in February, and Spain in March, to war with France. The fortunes of battle were varied. A Committee of Public Safety was instituted on March 25. The Girondins were expelled on June 2; Marat was assassinated on July 13; the Queen was executed on October 16, and the Girondins on October 31. But during these troubled months, Lavoisier was planning a scheme of national education for the Bureau of Consultation of Arts and Crafts.

His plan was presented to the National Convention in August. As many other works of this kind from his hand, it was published without the name of the author. The scheme devised related, first, to primary education and to the teaching of the useful arts. While recognizing differences in talent, it envisaged the complete democratization of opportunity. It advocated free primary education for all children, without distinction of social condition, "as a duty that Society owes to the child." Education at this stage, he thought, should include reading, writing, arithmetic, the elements of practical geometry, of botany, of natural history, and of agriculture; it should include, also, in the form of amusements and games, the art of working in wood and in metals. Secondary education should provide for two kinds of pupils, those intended for the public service, who would study languages and literature, and others destined for the mechanical arts. The former would be trained in schools resembling universities and colleges, and the latter in institutions still to be created, for girls as well as boys, and with no existing examples to copy. "There is no such example," said Lavoisier, "because no nation has been really interested in the most industrious class of the people." The need might best be satisfied, he thought, by the creation of primary schools of the useful arts in the chief towns of the various districts of France, with higher schools in the capital of each department for the most important subjects. In reality, he was presenting a plan for what did not then exist, but what we have now become familiar with—technical education; his projected institutions have been realized in our modern technical institutes and "polytechnics."

It was not merely the education of the individual that was to be considered, said Lavoisier, but the far greater aim of the education of the whole nation, of all mankind. For primary and secondary education he included drafts of the necessary decrees, including every detail of organization from the village school to the Faculties and learned Societies. He was to some extent inspired by the similar proposals made by Condorcet, but his scheme was more complete and more detailed. Some of these details are of more than passing interest: in the primary schools, lessons in one subject must not be long because prolonged attention to the same thing is tiring to young minds; the teaching of reading, writing, and arithmetic, must be interspersed with lessons on natural history, the structure of plants and animals, historical narratives, accounts of patriotism and benevolence, with walks relating to the arts and crafts of the countryside, and so on—all this part of the instruction being given as a diversion and a pleasure; a system of punishments should be drawn up and punishment given only after the teacher had called on a jury of children to decide whether a breach of rules had been committed or not.

In presenting the plan to the National Convention, it was made clear that its main purpose was to direct the attention of that body to education in the useful arts. But it added: "It must not be thought that it is possible to encourage some branches of the arts exclusively and to abandon others; the arts, the sciences, even literature, are bound by invisible links that cannot be broken with impunity; as it is chiefly from the sciences that the arts are enlightened, the same people cannot be great in the arts, if they are not at the same time great in the sciences. . . ." Moreover, it was not entirely a matter of spreading existing knowledge among as many individuals as possible, but also one of increasing knowledge itself. Here he advocated the establishment of institutions concerned with extending the limits of knowledge, with providing for industry new means of prosperity, and with insuring to the nation an increasing and lasting superiority in all its relations with other peoples. He had, indeed, envisaged the modern technical research institute.

Some of Lavoisier's ideas on this problem are still worth quoting in detail and they are still relevant to its solution, which is not yet complete everywhere:

> The aim would remain still unachieved if scientists and technologists [we use the modern term, which, although then uncoined, admirably expresses his meaning] charged with the advancement of human knowledge always dwelt in isolation from one another, or if they lived only as men bound to one and the same kind of study, industry or science; all branches of science and technology are linked with one another; it is not possible to make great progress in one, if all the others are backward; this is an army that must march on an even front. Moreover, most of the work still to be done in science and in the useful arts is precisely that which needs the collaboration and co-operation of many scientists. The geometer can well work alone at the perfection of the science of calculation; but what purpose is there in that, if the astronomer and the physicist do not continually provide him with observations and experiments to which it can be applied? Where would experimental

> physics and chemistry itself be, if geometry had not introduced its method and its severity into their reasoning and their calculations? It has already been observed that many of the useful arts, and those not the least important for Society, have need of collaboration from all the sciences; the art of navigation and the art of healing have been named; many others could be mentioned. This is why it is necessary for scientists and technologists to meet periodically in common assemblies and why these meetings should cover even branches of knowledge that seem to have the least relation and connection with one another.

The joint scientific and industrial societies and conferences of a later day!

> Declared Lavoisier:
>
> Citizen representatives! The fate of the French republic is in your hands. It rests only with you to raise France to a degree of splendor and prosperity higher than that attained by any nation whose memory has come down to us. Organize education everywhere; give impetus to the useful arts, the sciences, industry, commerce! See with what vigor all other nations, our rivals, are busying themselves with the means of supplying by industry what they lack in power, population, territorial wealth! A nation that does not take part in this general movement, a nation among whom the sciences and the useful arts languish in stagnation, will presently be outdistanced by rival nations; it will lose little by little all its means of competition; its trade, its power, its wealth, will pass into the hands of foreigners and it will finally become the prey of those who decide to invade it. Let a great empire, China, serve as a lesson for us; there the useful arts are today as they were two thousand years ago, because the form of government enchained the genius of the sciences, because it set limits that

industry cannot overstep. Legislators, education made the Revolution; let education still be amongst you the palladium of liberty! Now that you have completed your work, it only remains for you to animate it with the torch that you have in your hands.

The scheme proposed also twelve *Lycées nationaux,* or National High Schools, at twelve of the great cities of France, including Paris; all branches of knowledge would be studied in these institutions. At Paris also there would be a *Lycée central,* a special organization for higher learning based on existing institutions, so that literature would be studied at the *Bibliothèque nationale,* the national library of France, and astronomy at the Observatory. There would be four learned societies, replacing the former academies and similar societies; but all would be linked, as, in fact, they were afterwards linked by the creation of the *Institut de France*. The detail of every part of the proposal was completely worked out.

Lavoisier continued to take part in the meetings of the Bureau of Consultation right up to the time of his arrest. He was appointed President of the Bureau on October 26, 1793. His colleagues included many friends and associates from the Academy and the other learned societies, such as Borda, Vandermonde, Coulomb, Berthollet, Meusnier, Lagrange, Laplace, Hassenfratz, and others.

XXV

LAST DAYS OF THE ACADEMY

AFTER HIS ELECTION TO THE ACADEMY OF SCIENCES AS AN assistant in 1768, Lavoisier became an associate in 1772 and a *pensionnaire* in 1778. We have seen something of the active part that he took in all its activities and the keen interest he displayed in its internal affairs. His independence of spirit made him an alert defender of its privileges, particularly against the encroachments of its presidents who, nominated by the King from among the honorary members, derived neither their office nor their authority from the suffrages of their colleagues. When the Duke de la Vrillière, President in 1775, claimed that he had the right to decide without consulting the members what lectures should be given at the public meetings—for which it was known that he would choose some of his favorites, Lavoisier warned the Director, d'Arcy, of the danger of permitting this; it was, moreover, a distinction to be invited to give one of these lectures. Lavoisier wished also to extend the voting powers of the Academy so that members might choose their officers themselves instead of accepting the Minister's nominees. When he was himself appointed Director in 1785 and the Minister nominated him to the King, he asked (unsuccessfully however) that the members of the Academy be allowed to make the choice themselves.

As Director he discharged his duties conscientiously and admirably. He arranged with the President for the order of the lectures to be given at the public meetings; discussed with his colleagues the programs for the prizes offered for competition; endeavored to overcome the delay in the publication of the Academy's annual volume of *Memoirs;* arranged for the various committees to meet at his house; organized the preparations for special studies and the collection of necessary data; presented newly elected members to the King, and separately to the Queen and the other members of the Royal Family, and to the Ministers, as custom prescribed, with a whole detailed ceremonial. He had also to present, with appropriate formalities, the annual volumes of *Memoirs* of the Academy and books published by its members.

But it was to complete internal reform that he especially turned his mind and his efforts during his term of office as Director. He secured the abolition of the humiliating distinction of "assistant," a title introduced in 1716 to replace that of "pupil," which dated from the reorganization of 1699. "The grade of pupil," he said, "presents a humiliating distinction and is not suited to men of established reputation. . . . Its unsuitability was thought to have been remedied in 1716, when the title of *assistant* was substituted for that of *pupil;* but change of name has not changed the thing."

Other urgent changes were being considered, including an increase in the number of the sciences represented in the Academy. To the six already represented—geometry, astronomy, mechanics, medicine, chemistry, and botany —Lavoisier proposed to add three—physics, natural history, and agriculture. The Minister approved of this suggestion and the Academy was asked to appoint a committee to report on it. When Lavoisier's proposals were discussed by the committee, a new proposal was made, namely, to elect members for two further sciences instead of three, to fill these places with seven members in the different grades for each of the two sciences, and to elect eight agricultural associates. His reply to this, however, was: "To multiply the number of places in the Academy is to diminish the respect in which it is held, and this ill consequence is so much greater at the present time, when there has never been more pretension to knowledge and when there are, moreover, but few scientists who are not members of the Academy. You have many times experienced this yourselves, when it has been a question of replacing losses through death and, when vacancies occur, you often have a very small number to choose from. It is therefore not the case that scientists lack an Academy, but that the Academy is short of scientists; if, in these circumstances, you have to elect a large number of members, you will have no other resource but to call on mediocrity of talent, on the half-knowledge that is more dangerous than ignorance and the charlatanism and intrigue that go with it, and you will leave to succeeding generations only a stunted posterity, little worthy of that title of Academician which you have made illustrious."

The committee having thereupon rejected the latter proposal unanimously on April 13, 1785, Lavoisier succeeded in adding physics and natural history to the sciences represented; the grade of assistant was eliminated. The total number of places was thus increased from forty-two to forty-eight and, as there were already five supernumeraries, only one further nomination for a place was necessary. Lavoisier failed, however, in his struggle to secure for the associates the right to vote in the affairs of the Academy and thereby to achieve equality among the different grades. When the reforms were approved by the Minister, discontent in the minds of some members had not been allayed, and it broke out again when the division of the sums assigned by the Government for the salaries of *pensionnaires* and associates had to be adjusted in accordance with the changes. There had always been an order of precedence in this matter: a member passed successively through the grades of third associate, second associate, and first associate, then to third *pensionnaire,* second *pensionnaire,* and so on. Rearrangement presented problems; the new places in physics and natural history had to be filled from among the members, the six assistants had to be fitted in as associates, and so on. When the proposed adjustments were announced, Lavoisier had to hear many complaints from those who felt injury to either their self-respect or their interests. But the list was revised and Lavoisier was able to announce in a letter to the Vice-President that almost every member had gained by the revision and that satisfaction was practically general.

We have already described the scientific labors of Lavoisier in the Academy, the great discoveries unfolded in the classic memoirs that he read in its assemblies, and the numerous valuable reports that he compiled for its committees. It is appropriate now to give some brief account of the last researches that he described to his fellow members in this ancient society.

We have seen (Chap. XIII) that, in experiments with Laplace, Lavoisier had concluded that respiration was a kind of slow combustion, and that the maintenance of the animal organism at a constant temperature above that of its surroundings was due to the heat liberated in respiration from the oxygen of the air by its conversion into fixed air through its combination with the base of fixed air (carbon) in the blood. In these experiments it was found that the heat evolved by the formation from oxygen and carbon of an amount of fixed air equal to that produced by the respiration of a guinea pig during 10 hours melted 10½ ounces of ice, while the heat lost by the animal during the same time melted 13 ounces of ice. It was, therefore, concluded that, allowance being made for the unavoidable error in such experiments, these figures showed sufficient agreement to justify the conclusion that animal heat was maintained at a constant temperature above that of its surroundings by the conversion of oxygen into fixed air in the process of respiration.

In 1785 Lavoisier reported to the Royal Society of Medicine that in some new experiments he had found that the oxygen consumed in respiration was not completely accounted for by the amount of fixed air produced. He concluded that the missing oxygen had combined either with the blood or with hydrogen in the lungs to form water; and he preferred the latter explanation, although, of course, he made no suggestion about the source of the hydrogen.

Presently Lavoisier returned to this problem and took as his assistant, Séguin, then aged about twenty-one. Their first experimental results were announced to the Academy on November 13, 1789. In their introduction Lavoisier made it clear that he no longer regarded the discrepancy in the values obtained with Laplace some years earlier as due to experimental error. He now concluded that these differences showed that in respiration more heat was liberated from the oxygen than would be produced by the formation of the fixed air alone. The liberation of this further amount of heat was due to the formation of water through a combination of an additional amount of the oxygen of the air with hydrogen contained in the blood. Animal heat was maintained by this dual conversion in the lungs during respiration, a part of the oxygen of the air combining with carbon in the blood to form carbonic acid gas and another part combining with hydrogen in the blood to form water.

With guineapigs as subjects, they now showed in quantitative experiments that the amount of oxygen consumed in respiration during a fixed time was the same whether oxygen or air was respired and that neither nitrogen nor hydrogen, added to the oxygen breathed by these animals, was in any way changed in respiration. Having obtained some evidence that the amount of oxygen consumed was greater during digestion or movement, they decided on a direct study of human respiration with Séguin as subject.

In these further experiments, they found: (1) that a fasting man at rest consumed more oxygen in cold than in warm conditions; (2) that during digestion the rate of oxygen consumption increased to about half as much again; (3) that during exercise or movement, or, in other words, during work, the amount of oxygen consumed increased to nearly three times the amount consumed by a fasting man at rest (Séguin, for the purposes of this experiment, constantly lifted a weight of 15 pounds during the quarter of an hour during which his rate of consumption of oxygen was measured. The work that he performed was actually equivalent to the raising of 15 pounds through a vertical height of 650 feet); (4) that, when the same work was done during digestion, the amount of oxygen consumed was nearly quadrupled; and (5) that, when the rate of consumption of oxygen was increased in these ways, the pulse rate increased also, while throughout the tests the temperature of the subject remained practically constant, variations amounting to only a fraction of a degree. Their results showed, therefore, that the rate of consumption of oxygen in the lungs depended upon the temperature of the air, being greater on a cold day, and that it increased during digestion and during the performance of work. These important results marked a striking advance towards modern physiology, although later discoveries have shown that the chemical process and detailed mechanism of respiration are more complicated.

To Lavoisier, the mechanism of the animal body was under the control of three processes, respiration, transpiration, and digestion. In respiration, some carbon and hydrogen were consumed in the lungs by combination with the oxygen of the respired air, the oxygen giving up its matter of heat, which was then carried round by the circulating blood and served to maintain the body at its proper temperature above that of its surroundings. In transpiration, that is, in the production of insensible aqueous perspiration, occurring on the surface of the skin and in the lungs, water was exuded and evaporated, and its evaporation absorbed matter of heat in the proportion requisite to prevent the temperature of the body from rising above its proper level. In digestion, food restored to the blood what it had lost in respiration and transpiration.

Reflecting on his view that the ultimate source of the heat of the animal body was the air, Lavoisier remarked: "One would say that the analogy between respiration and combustion had not escaped the poets, or rather the philosophers of antiquity whose interpreters they were. The fire stolen from heaven, the fire of Prometheus, is not merely an ingenious poetical idea; it is a faithful picture of the operations of Nature, at least for animals that breathe: we can therefore say with the ancients that the flame of life is lit at the instant when the child draws its first breath and that it is extinguished only at death. When we consider these remarkable anticipations, we are sometimes tempted to think that the ancients actually penetrated further than we suppose into the sanctuary of knowledge and that the fable is, indeed, only an allegory under which they hid the great truths of medicine and physics."

There were other considerations, too, in his mind and he reached out towards our modern notions of energy, commenting on his general results: "Observations of this kind lead us to compare different expenditures of strength between which there would seem to be no relationship. We could ascertain, for instance, the number of pounds, in weight, that are equivalent to the efforts of a man who delivers an oration or of a musician who plays an instrument. We could even evaluate the mechanical effort in the work of the philosopher when he is reflecting, of the man of letters when he is writing, of the musician when he is composing. These tasks, considered to be purely intellectual, have nevertheless in them something physical and material which allows us, according to the relationship that we have established, to compare them to those of the laboring man. It is therefore not without reason that the French language has confounded under the common denomination of *travail* the efforts of the mind with those of the body, the *travail* of the study and the *travail* of the hired servant."

And there were social considerations, too, in expressing which Lavoisier voiced his characteristic Liberalism, derived in France from the work of the *philosophes* and, unfortunately, soon to be submerged in the bloody tide of revolution:

So far as respiration is considered to be mere consumption of air, the lot of the rich and the poor is the same; for the air belongs equally to all and costs nothing: the laboring man who works more enjoys even more fully this gift of Nature. But now that experiment has taught us that respiration is a real combustion that consumes at every instant a part of the substance of the individual, that this consumption is so much greater as the circulation and respiration are accelerated, and that it increases proportionately as the individual leads a more laborious and more active life, a host of unforeseen ethical considerations arises from these physical results. By what mischance does it come to pass that the poor man, who lives by the sweat of his brow and who is compelled to put forth for his maintenance all the strength that Nature has given to him, expends more than the leisured man, who has less need to recruit his energy? Why, by a shocking contrast, does the rich man enjoy an abundance which is not physically necessary to him and which seems destined for the laboring man? Let us beware, however, of calumniating Nature and of accusing her of faults that, no doubt, originate in our social institutions and are perhaps inseparable from them. Let us be content with blessing the philosophy and the humanity that have now joined forces to promise us wise institutions which will tend to equalize wealth, to raise the price of work and to insure its just reward, and to offer to all classes of society, and especially to the poor, more enjoyment and happiness. Let us resolve above all that the enthusiasm and exaggeration that so readily possess men united in crowded assemblies, and the human passions that carry away the multitude so often against their real interest and include in their surge the sage and the philosopher with other men, shall not wreck a project so promisingly begun and destroy the hopes of our country.

Finally, he had a word to ambitious politicians on the services of science to humanity: "We shall end this memoir with a comforting reflection. It is not absolutely necessary, in order to deserve well of mankind and to do one's

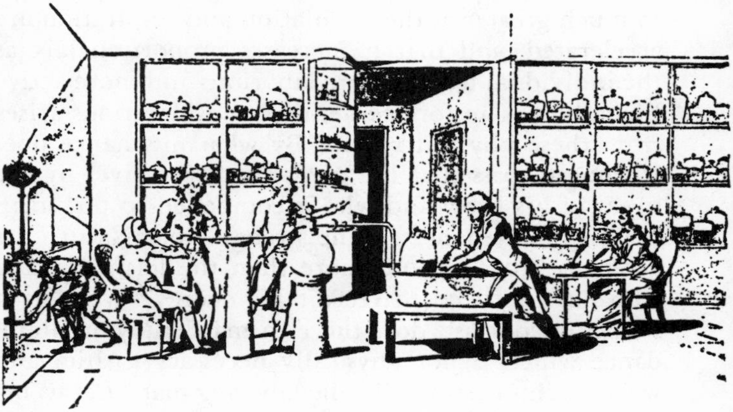

26. Experiments on the respiration of a fasting man at rest made in Lavoisier's laboratory at the Arsenal *c.* 1789. The subject is Séguin, who is breathing oxygen, supplied from a receiver in a trough, through a mouthpiece fitted to a mask enclosing his head and devised for these experiments. Mme. Lavoisier, who made this sketch, is writing the laboratory record.

duty to one's country, to be called to glittering public office for the organization and regeneration of empires. The man of science in the silence of his laboratory and his study can also serve his country: by his work he is enabled to hope that he may diminish the sum of the evils that afflict the human race, and increase enjoyment and happiness; and were he to contribute, by the charting of new ways, only to the prolongation by some years, even by some days, of the average life of men, he could aspire also to the glorious title of benefactor of humanity."

The researches on respiration and transpiration were continued and further accounts were given to the Academy

27. Experiments (*c.* 1789) on the respiration of a subject (Séguin) doing work, moving a weight by means of a treadle. Oxygen is supplied as in the preceding illustration. The sketch was made by Mme. Lavoisier, who is shown writing the laboratory record. The apparatus on the right is a Ramsden electrical machine.

of the progress of the experiments during 1790, 1791, and 1792, but Lavoisier had other pressing duties and the work was eventually interrupted by his death. These great researches were never completed. The most important facts, however, were set out in the memoir quoted above. Sketches made by Mme. Lavoisier of the experiments on respiration in progress in the laboratory are reproduced in Figs. 26 and 27.

Lavoisier was, moreover, busy with other scientific matters, especially with the work on the new system of weights and measures, which eventually developed into the metric or decimal system. France had always suffered from an extraordinary diversity of weights and measures. Lavoisier and others had long urged the introduction of a uniform system for the whole country, while some of the *cahiers* sent by the Provincial Assemblies to the States-General in 1789, including as we have seen that of the nobility of Blois, had recommended such a reform. By a decree of May 8, 1790, on the proposal of Talleyrand, the National Assembly entrusted to the Academy of Sciences the duty of formulating a system founded on unchanging bases and capable of adoption by all nations.

A Commission was appointed by the Academy and sub-committees were appointed later for various necessary tasks. The Commission, consisting of Laplace, Lagrange, Borda, Monge, and Condorcet, reported to the Academy on March 19, 1791, that the basis of length should be related, not to the length of a pendulum beating seconds, since this length varied with latitude, but to the length of a quadrant of the earth, a quarter of the great circle passing round the earth through the poles, the ten-millionth of which should constitute the practical unit of length, since known as the meter. It announced the composition of the sub-committees, to one of which Lavoisier was appointed with Haüy to determine the weight of a volume of distilled water in a vacuum at zero temperature on the thermometric scale. From this determination the unit of weight, the gram, would be derived, as the weight of distilled water occupying a cube, each side of which measured one centimeter, that is, one-hundredth of the unit of length, the meter.

Further, Lavoisier was appointed secretary of the Commission and treasurer, as a sum of 300,000 livres had been voted by the Assembly for the expenses likely to be incurred by the Academy in the course of the work. Instruments had to be prepared and measurements were not begun until 1792. Méchain left for Spain in June and Delambre went to Dunkirk in order to measure the distance between two points on the earth's surface and thence calculate the length of the earth's quadrant. Many difficulties were encountered, especially after the outbreak of war, when Lavoisier had to contrive to pass money to Méchain through Switzerland. During the progress of the work all reports were sent to Lavoisier as secretary and presently he was appointed to succeed Laplace when the latter withdrew from the Commission.

In sending to Loysel, the representative of the Assembly presiding over the committee on *assignats* and coinage, a memoir on the measures of older times, Lavoisier suggested that money also should be rearranged on a decimal system, and included with his suggestion a draft of the necessary decree. For the manufacture of the standards to be used in the new system of weights and measures, he applied to the Committee of Public Instruction so that the craftsmen engaged might be exempted from military service. He had to make innumerable applications to the Treasury to obtain the funds voted by the Assembly for the work of the Commission, the history of which is so closely linked with the history of the Academy in its last years, and with its fate.

Lavoisier had become treasurer of the Academy in December, 1791. Almost immediately there were many difficulties, some caused by the long illness of his predecessor and others arising from the growing hostility to all institutions that survived from the old regime. "Lavoisier," wrote Grimaux, "devoted himself passionately to the defense of the learned society; for more than twenty months, he was the very soul of the dying Academy. To save it, to save the sciences of which the Academy was the protector, he deployed all the forces of his prodigious energy; without faltering for an instant, he fought to the last hour before he was beaten in this unequal contest in which he stood almost alone against the will of the Convention, which made a clean sweep of the institutions of the past, even the most inoffensive of them. His courageous intervention at a time when perhaps it would have been wiser for him to remain in the shadows, his generous devotion to science, his patriotic concern for national honor, make him great as much as do the discoveries to which he owes the immortality of his name."

A few days after his appointment as treasurer he declined the 3,000 livres expense allowance. He reminded the Academy that the law and especially public opinion, most powerful of all laws, did not permit him to have two salaries at the same time and he was still a Commissioner for Gunpowder. He proposed to allot 1,000 to 1,200 livres for the clerks in the treasurer's office and for office expenses, and to put the remainder at the disposal of the Academy.

As a result of the long illness of Tillet, whom he had succeeded as treasurer, much had been left undone. Salaries of the Academicians were in arrears; sums due to the Academy for 1790 and for 1791 had not been received from the Treasury. The Minister had to be approached and urged to expedite the order for payment. The administrative office for Paris had to be supplied with a statement on the finances of the Academy, an inventory of its possessions, and a list of its members. Again, the Minister of Marine had not paid the annual subvention of 1,600 livres for the work on the nautical almanac (which it was the duty of the Academy to publish), even though Méchain had completed the calculations for 1793 and was working on those for 1794. When the Minister was finally ready to put the matter in order, the secretary of the Academy, Condorcet, had to make formal application for payment; but Condorcet was otherwise occupied. Lavoisier wrote the letter for Condorcet's signature. Even on comparatively trivial matters, it seems to have fallen to Lavoisier's lot to defend the interests of the Academy—for instance, when the Keeper of Archives asked the Convention for authority to make use of the attics of the old Louvre above the Academy's meeting rooms, where they had stored their apparatus for the last ten years, Lavoisier had to reclaim possession from Roland, Minister of the Interior.

It was a period when every member of the Academy leaned upon him. Sage, bitter opponent of the new chemistry, did not hesitate to borrow from him 1,800 livres repayable on receipt of his salary. When the rent of 500 livres, due for Lemonnier's observatory and normally paid by the State, had not been received after repeated applications, Lavoisier obtained it from the Academy. Lemonnier was now seventy-seven, ill and paralysed; he was the last survivor of the expedition sent to the Arctic in 1736 to measure a degree of the meridian for the calculation of the earth's dimensions.

All the duties were exacting, including reports on the finances of the Academy, interviews with Ministers, the work on weights and measures, and requests to officials for payment of the subventions voted by the Assembly. Of the 300,000 livres voted for the Commission of Weights and Measures, 100,000 had been paid for 1791. A similar sum was due for 1792, but when Méchain and Delambre were on the point of departure, it had not been received, despite Lavoisier's applications to the Minister; in September, when Lavoisier was insisting on immediate payment, the two astronomers had already begun their work. It was probably at this time that he became too occupied with other matters to pursue his experiments on respiration.

Life, however, went on in the Academy; meetings were regularly held and important tasks were on hand. As the year 1792 opened, all that might have been noticed as different from what it had been in other days was the absence of the honorary members, almost all of whom were nobles and had either crossed the frontier as *émigrés* or were living far from Paris. But at the meeting of April 25 conditions explosively changed. Fourcroy invited the Academy to follow the example of the Royal Society of Medicine, which had removed from its list of members the names of many of those who had left France as *émigrés* and those well known to be "counter-revolutionaries." He proposed that the list of members be read so that the names of certain of them known for their *incivisme* might be deleted. Every man felt the menace of these words; it was clearly not the mere proscription of *émigrés* that Fourcroy had in mind. Several members objected that the Academy had no jurisdiction over the political views of its members and that the pursuit of science was its sole occupation. Fourcroy, however, persisted in his proposal, but consideration was postponed until April 29. On that date a motion was carried to submit a list of the members to the Minister of the Interior who would order dismissals, if there were to be any, while the Academy "would apply themselves, as was their custom, to more intellectual occupations." Fourcroy did not forget the humiliation of this defeat and the Academy later felt his resentment; enthusiastic supporter of the Revolution, he became the most bitter and inflexible of the Academy's enemies.

Lavoisier returned to Paris from what proved to be his last visit to Fréchines to be present at the opening meeting of the new session of the Academy on November 14. On November 25, the Academy, in the hope of securing support from the National Convention, appeared in a body at the bar of the house to present a copy of their annual volume of *Memoirs* and a report on the progress of their work on the new system of weights and measures. Borda, entrusted with the making of the report, expressed the hope that all would be complete early in 1794. "Then," he said, "it will only remain to make the standards that are to be sent to the different nations and, perhaps, also to the learned societies of Europe who, by their reputation, can contribute most in extending their use: the Academy will consider itself fortunate in having been able to contribute to this work by its own efforts and it will ever congratulate itself on having co-operated in the execution of a project, which redounds to the glory of the nation and which is useful to all mankind and capable of becoming a new link of universal brotherhood among all peoples who adopt it."

The Academicians had made a favorable impression, and Grégoire, the President, replied to their greeting and to Borda's report:

> Citizens, the National Convention commends the importance and the success of your work. For long the philosophically minded have included in their aims the emancipation of men from the varieties of weights and measures, which have ever been a hindrance in their transactions. . . . It is through you, worthy men of science, that the world will owe this gift to France. You have derived your theory from Nature; from all determinate lengths, you have chosen those two alone that in combination give the most absolute

result, the measurement of the pendulum and especially the measurement of the meridian. . . . Yours will be the glory of having revealed this enduring unity to the whole world; this beneficent truth, which shall forge a new bond between nations, is one of the most useful victories of equality. The National Convention accepts the precious collection that you have presented to it and invites you to its meeting.

Although the Convention passed a decree that Borda's report and the President's reply should be printed, three days later another decree was passed which prohibited the Academy from making appointments to vacant places until further orders. The flattering reception could not conceal the presence of hostile elements in the Convention.

During this winter of 1792 Lavoisier worked with Haüy on the determination of the density of pure water for the Commission of Weights and Measures. As treasurer of the Academy, he had to remind Monge, now Minister of Marine, about the subvention due from his department for the publication of the nautical almanac, writing: "The Academy considers itself fortunate in having on this occasion an interpreter at the National Convention who adds to the double qualification of scientist and Academician that of Minister of the Republic." Publication of the annual volume of *Memoirs* had to be seen to. Likewise, accommodation for a chemical laboratory in the Louvre had to be discussed with Roland, Minister of the Interior. Roland had returned the revised list of members of the Academy, a communication which seemed to afford a kind of official recognition of the Academy, but which, however, proved of no advantage in the end.

The decree forbidding any appointments to vacant places until further orders cast a gloom over the affairs of the Academy. Minutes of the meetings of 1793, says Grimaux, do not give the names of members present and do not state what subjects were discussed, but indicate merely the work done and the reports compiled at the request of Ministers and of the Convention. The Ministers of War, Marine, and Public Revenues, referred one problem after another to the Academy, including military hygiene, armaments, water supply, the savings bank, and a scheme of regulations for the delivery of saltpeter, described as "due to the devotion of Citizen Lavoisier," who had embarked on it when he was still Commissioner for Gunpowder. The Government was always ready to make full use of the services that the Academy and the Academy alone could render. It was, however, Lavoisier who kept the Academy alive and it seemed that he might yet be able to save it.

"All was, however, crumbling about him," wrote Grimaux; "Louis XVI, who a year ago had offered him a Ministry, had just died on the scaffold; the wealthy and elegant society, interested in the things of the spirit, that had welcomed his labors, had disappeared in the storm; while the world that he had known went down to disaster, he remained the vigilant protector of the higher interests of science; he did not intend that science should disappear from the soil of France where it had flourished so magnificently."

Research must not cease; money for one purpose after another was obtained. When the salaries of the Academicians were not paid, Lavoisier applied to the numerous officials who dealt with this matter. When the revenue office refused payment in virtue of the decree suppressing all corporations until a new decision had been made, he warned the Academicians who had been elected to the Convention, especially the geometer Arbogast: "Foreign powers ask nothing better than to turn this circumstance to profit by transplanting the sciences and the useful arts to their own territories, but it can be said in favor of French scientists that their attachment to their country has been unshakeable and that there is not one of them who is not ready to reject with indignation all such proposals that might be made to him."

Every possible step was taken by members of the Academy to avoid any appearance of lack of sympathy with the Revolution. In April they contributed a "patriotic gift" of 11,845 livres to the National Treasury and presented also a piece of gold received by bequest and valued at 12,000 livres. Later they ordered the removal of the tapestries from the room in which they held their meetings in the Louvre, "because these tapestries display symbols that should be banished from a republican country."

The question of the salaries of the Academicians now stood referred to the Committee of Public Instruction and on April 28 Lakanal was appointed arbitrator on the matter. Lakanal was young and generous and proved the last friend of the Academy. To him Lavoisier wrote promptly, showing that the decree of March 8 governing payments to institutions concerned with public instruction was applicable to the Academy and adding: "Never has the Academy been charged with more numerous and more important labors for the common weal: you may judge for yourself from the attached statement. These tasks cost the nation nothing and the members of the Academy are all the more zealous in carrying them out because they are gratuitous. Is it just, when they are overwhelmed with work, to dispute the means of existence which they have a right to expect in accordance with the decrees of the legislature? Citizen, time presses; many of the Academicians suffer; several have already left Paris because their means no longer allow them to live there; science, if no one comes to the rescue, will fall into a state of decay from which recovery will be difficult."

Lakanal fought for the Academy; and he was at first successful. He obtained on May 17 a decree from the Convention permitting the members to make appointments to vacancies; and on May 25 he obtained another, by which the salaries of the Academicians were to be paid as in the past, with some minor conditions, such as certified residence of six months in the territory of the Republic and a statement from the Minister. But this second decree was won only after a hard struggle, about which Lakanal later said: "The task that I set myself in the cause of the Academy of Sciences was very difficult. The Committee of Finance, otherwise composed of good citizens, was intractable when I asked for funds for the scientists, the Academicians. What rebuffs I met with! It was popularly believed that they were all opposed to the new state of things, and unfortunately there was some truth in this supposition. A course had to be steered through all these dangers."

Administrative formalities, however, made the decree of May 25 all but illusory. In accordance with the conditions prescribed, Lavoisier obtained the necessary documents, but on July 10 no money had been paid over by the National Treasury. Once again, as in 1792, he advanced the salaries from his own fortune. But the decrees passed through Lakanal's generous support gave the Academy some reassurance; research continued and reports on problems were made at the request of the Government. One of these was on the best means of preserving ship's biscuit, in which the Minister concerned wrote: "I look for this office to a Society that has always regarded the amelioration of man's estate as the most valuable service that learning can render." As late as August 1, survival of the

Academy seemed assured. On that day, in passing the decree for uniform weights and measures based on the meridian and the decimal system, the Convention congratulated the Academy, directed them to watch over the execution of the decree, and presently sought their advice on the determination of the fineness of the gold and silver coinage.

Appearances were deceptive: on August 8, 1793, after a report by Grégoire, submitted by the Committee of Public Instruction, the Convention decreed the suppression of all learned societies.

"The Convention," wrote Grimaux, "bears the entire responsibility for the suppression of the Academy of Sciences." The report of Grégoire, while sharing the general opinion in favor of the suppression of the academies, was intended to exempt the Academy of Sciences and also to allow the dissolved societies to reconstitute themselves on new bases. After stating that the *Académie française,* the literary academy, showed all the symptoms of decay and that the Society of Medicine, the Society of Agriculture, and the Academy of Sciences, had displayed more energy since the Revolution, Grégoire added:

> The Academy of Sciences, which has always been composed of the first scientists of Europe, has described more than 400 inventions and published 130 volumes which form one of the greatest monuments of the human mind; it continues with admirable activity the tasks that you have entrusted to it on the silver of the suppressed churches, on the fineness of the gold and silver coinage, on the production of saltpeter, and on the measurement of a degree of the meridian, a task that will need a year for its completion. You have just adopted its work on weights and measures; it is engaged on the construction of the new standards and of the comparison of the new measures with all those

> that have been hitherto used in the different provinces of France. . . . Academicians devoted to the cultivation of the sciences can scarcely be reproached for that *esprit de corps* which is to societies what egoism is to individuals. . . . We should be dishonored if our men of science were reduced to carrying to foreign shores their talents and our shame. . . . Observe how all the useful arts, all the sciences, go hand in hand, from the plane to the astronomical laws of Kepler, to the most abstract profundities of the calculus and of astro-physics, by which our scientists have increased the sum of knowledge that we owe to the genius of Newton, to the most sublime researches in chemistry, for again it is the Academy of Sciences that has regenerated chemistry and presented to an astonished Europe the only theory that accords with Nature.

The draft of the decree contained seven clauses; the first suppressed all academies and literary societies, but the remaining clauses provided for exceptions and safeguarded the interests of science. It stated that: "The Academy of Sciences remains provisionally charged with the various tasks entrusted to it by the National Convention; it will therefore continue to receive the annual grants made to it until otherwise arranged; the National Convention directs its Committee of Public Instruction to submit without delay a scheme for the organization of a society for the advancement of science and the useful arts. . . . Citizens are entitled to combine in free societies to contribute to the progress of human knowledge."

The decree had been drafted by Grégoire on the advice of Lavoisier. It would have satisfied the current feeling for reform and saved science and learning in France, but for a venomous speech by the painter David, who, however, said that he was alluding only to the Academy of Painting and Sculpture. But he added: "To speak of one Academy is to speak of all. . . . In the name of justice, for love of art, and above all for our love of youth, let us destroy, let us annihilate these baneful Academies, which cannot exist under a free government." The Convention was carried away by David's oratory. It passed only the first and last articles of the decree submitted by the Committee of Public Instruction, the former suppressing all academies and the latter putting their property in the care of the State.

Lavoisier was not yet defeated. He wrote to Delambre, then at Amiens, to continue his work on the measurement of the meridian; and then he sought out Lakanal to propose the continuation of the work of the Academy through such a new free society or club as had been suggested in Grégoire's report and authorized by the Constitution. Meanwhile, the Academy held its last meeting on June 29; the page from the records with the signatures of those present is reproduced in Plate 28. Several members presented memoirs which they hoped would be included in the last number of the long series of 146 volumes that had appeared since the original foundation of 1666. Although official notification of the decree of suppression had not yet been received, it was decided to set an example of obedience to the law and therefore to hold no more meetings.

Lavoisier now turned to Lakanal and the Committee of Public Instruction. Opening his letter with "now that the Academy is no more and since the motives that have led to its suppression have no connection either with its administration or with its work . . . ," he went on to point out that the decree of August 8 meant that instruments would now have to be taken from the hands of those using them,

…ont assisté à la Séance du Vendredy 29 Juin 1793
Méssieurs les Académiciens soussignés

Méssieurs

sont arrivés après l'heure

Méssieurs

28. The signatures recorded at the last meeting of the Academy of Sciences on June 29, 1793, before its suppression on August 8, 1793. The names read, in order, Lemonnier, the Abbé Bossut, Desfontaines, Desmarest, Pingré, Baumé, Adanson, Brisson, Tenon, d'Arcet, the Abbé Haüy, Jeaurat, Lagrange, Demours, Messier, L'Héritier, Lamarck, Daubenton, Fourcroy, de Jussieu, Bory, Sabatier, Lavoisier, Lalande, Buache, Le Roy, Berthollet, Monge, Le Gendre, Borda, Vic d'Azyr, Vandermonde, Cassini. *By courtesy of the Académie des Sciences from the original.*

publication of important books and of the sheets of the geological atlas would have to be abandoned, researches that had been begun would have to cease, and the continuation of the work on weights and measures would be left in doubt, while the apparatus for it, the balances and the standards, were in course of manufacture and accounts had been opened with the makers. To avoid these disasters, he suggested that the members of the Academy might be allowed to form a society for the advancement of science with the moneys allocated to the suppressed Academy and under the control of the Convention.

He pleaded for his colleagues, the withdrawal of whose salaries was placing them in difficulties:

> It is needless, citizens, for me to add that the continuation of salaries to those who have obtained them is strict justice; there is no Academician who, if he had applied his intelligence and his abilities in other fields, would not have gained a place and a livelihood in society. Relying on public faith, they have followed a career which is doubtless honorable but of little profit. Several are over eighty and infirm; several have exhausted their strength and their health in journeys and labors freely undertaken for the government; French loyalty does not allow the nation to belie their expectations and they have at least a strict right to the pensions decreed in favor of all public officials. . . . Citizens, time presses; if you leave the members of the former Academy of Sciences time to withdraw to the country, to take other positions in society, to give themselves to profitable occupations, the organization of science will be destroyed and a half-century will not be long enough to produce another generation of scientists. For mercy's sake, for the honor of the nation, for the interest of society, for the opinion

> of foreign nations who have their eyes upon you, obtain a provisional decree to prevent the fall of the useful arts which would be the necessary result of the annihilation of science.

Next day Lavoisier sent another letter to Lakanal to propose that the Committee of Public Instruction should enter into relations with the new society that he had suggested for the advancement of science, "The Free Society for the Advancement of Science," and so give it a kind of recognition, which would relieve the anxieties of the Academicians. At the same time, he approached Arbogast again, pointing out the results of the suppression of the Academy on the work on weights and measures and stating: "I am in need, citizen, of immediate authorizations for these different objects, the more so as the funds that remain in my hands are quite insufficient to supply the needs of Citizens Delambre and Méchain and to settle the accounts of the craftsmen who have worked on the making of the weights and measures. If the Committee of Public Instruction does not obtain an immediate decision from the Convention on these different points, the work on the weights and measures will be completely suspended."

Arbogast was one of the two commissioners appointed for the weights and measures. The other was Fourcroy, an old pupil and associate of Lavoisier. Fourcroy had just been appointed, but it does not appear that Lavoisier made any approach to him.

Lakanal, in response to Lavoisier's appeal, and without the previous reference to the Committee of Public Instruction prescribed by formality, tried to secure a reversal of the Convention's hasty decision. On August 14, as a result of his efforts, the Convention decreed "that the members of the former Academy of Sciences shall continue to meet in their usual place of assembly to occupy themselves with the problems that have been and will be referred to them by the National Convention. . . . The annual sums paid to the scientists who composed the former Academy shall be paid to them as in the past."

The Academy seemed to have been saved; but Lavoisier's final disappointment was bitter. He promptly convened a meeting of his colleagues to be held on August 17 in virtue of the decree of August 14. When they arrived at the Louvre, they found that seals had been placed on all the Academy's rooms that morning by order of the Paris administration under the decree of August 8, the decree of August 14 being ignored. Lavoisier immediately called on Lakanal, who sent him, on September 1, a certified copy of the decree of August 14. It, however, remained inoperative, since both the Convention and the Committee of Public Instruction were hostile to the Academy.

In writing to Lakanal to thank him for his devotion to the Academy, Lavoisier said:

> Citizen representative, I have received, with feelings of gratitude that it would be difficult for me to express to you, the copy of the decree of the National Convention of August 14 that you have been kind enough to send to me. I have communicated it to some of my former colleagues. Unfortunately, circumstances do not appear to allow us to take advantage of this decree, however important it may be for the work on weights and measures and for the continuation of the other studies with which the Academy was entrusted by the Convention; as for assembling as a free society, we could not do so at present without giving the appearance of resisting the prevailing opinion of the Committee of Public Instruction and of the majority of the Assembly. . . .

LAST DAYS OF THE ACADEMY

It seem inexplicable that the Academy of Sciences was suppressed, especially after the Convention had passed the amending decree of August 14. In Grimaux's opinion, this decree remained ineffective through the bitter hostility of Fourcroy, who was not only a representative in the National Convention, but also one of the two commissioners appointed by the Convention for the work on weights and measures and recently appointed a member of the Committee of Public Instruction. Fourcroy had been rebuffed in 1792 when he proposed a "purifying scrutiny" of the membership list of the Academy; and later he had applied to the Academy, apparently without success, for a grant-in-aid for his researches. Grimaux states that Fourcroy was now at the head of a group in the Committee of Public Instruction which, either from indifference or hostility, ignored the decree of August 14 and concerned themselves only with the continuation of the work on weights and measures, which they entrusted to a special committee directed to carry on the work that the Academy had been doing on this problem. When he wrote to Lavoisier on August 20 to ask for a report on the state of the work of the committee appointed by the Academy on weights and measures and for a list of the scientists engaged upon it, his letter was couched in cold official terms.

The final act in the drama occurred at the meeting of the Committee of Public Instruction held on September 7, for which the minutes record: "September 7, 1793. Fourcroy, in the name of the committee directed to examine what work of the former Academy it was important to continue, after showing the necessity of prosecuting without delay the work on the weights and measures, proposed the draft of a decree in 7 clauses. Lakanal opposed the draft and moved the previous question, basing his objection on the fact that the decree passed on August 14 by the Convention maintained at their work and with their salaries the scientists of the former Academy entrusted with a great number of tasks by the Convention. Several members spoke in support, some for the previous question, proposed by Lakanal, others for the draft of the decree submitted by Fourcroy." Discussion went on until the meeting of the 9th, at which the Committee adopted Fourcroy's proposal for the setting up of a temporary commission of weights and measures. As Grimaux commented, this was the final sentence on the Academy.

The new commission was established by the Convention under the superintendence of Fourcroy and Arbogast. Its members included Borda, Laplace, Lagrange, Brisson, Haüy, Coulomb, Delambre, Monge, and Lavoisier. Once again the duties of secretary and treasurer fell to Lavoisier, who at once threw himself into the work of the new commission with all his customary energy and enthusiasm.

But already the shadows had begun to darken about him. On September 10 he was the subject of a domiciliary visit by the officials of the local section of the revolutionary committee and, indeed, only two months of liberty still remained for him.

XXVI

TRAGEDY

AFTER THE TRIAL OF LOUIS XVI AND HIS EXECUTION ON January 21, 1793, the Convention again turned its attention to the delay in the presentation of the accounts of the Tax-Farm, which had been suppressed on March 20, 1791. The preparation of the accounts and liquidation of the Farm's assets had been consigned to a commission; but the difficulty of finding time for these tasks while having to carry out their normal current official duties, such as the sale of tobacco and salt in store and the collection of customs dues, prevented the commission from completing their work on the date originally fixed by the National Assembly, January 1, 1793. Clavière, the Minister of Public Revenues, was well aware of their difficulties and at length relieved them of all duties except liquidation, reporting favorably on their zeal and their loyalty to the National Convention on December 31, 1792.

Popular imagination, however, still credited the Farmers-General with vast and illegally acquired fortunes, perhaps of 300 to 400 million livres, which would be very useful in filling the empty coffers of the State. Popular imagination was somewhat wide of the mark; Mollien, who had been attached as a government official to the Farm's office for some years, afterwards estimated that the Farm had lost 80 million livres in the general bankruptcy and that the Farmers-General as a body would have had difficulty in finding 22 millions by the sale of all their possessions.

On February 26, 1793, a proposal was made in the National Convention for the appointment of a new commission to collect information about the crimes, offences, and abuses, said to have been committed by the Farm against the State, and to extend these inquiries as far back as 1740. The sinister suggestion appeared that this must be done quickly because the Farmers-General, "the bloodsuckers of the people," were hastening to sell their property and possessions, not merely to avoid restitution, but also to take their ill-gotten gains abroad to the enemies of the Republic. Despite this more serious insinuation of *incivisme,* no decision was made by the Convention, which was then the scene of a fierce struggle between the Mountain and the Girondins.

But on June 5 following, the commissioners appointed to liquidate the Farm's assets and present the accounts were accused by Montaut, a representative in the Convention, of intentional delay in their work and of misappropriation of moneys entrusted to them. A decree was passed to suppress the commission, to seal their papers, and to confiscate the balance of their funds. All protests notwithstanding, the work of the commission had to cease. Once again, as with the decrees relating to the Academy of Sciences, these decisions were taken by the Convention as a result of personal intervention by a member without reference to the relevant committees. And some months later another decree ordered the sealing of the personal papers of the Farmers-General.

Lavoisier was now living at 243 Boulevard de la Madeline and on September 10 and 11, at the very moment when he had been entrusted by the Committee of Public Instruction with the organization of the new commission on weights and measures, he was subjected to a domiciliary visit by two representatives of the local section of the revolutionary committee, whose duty was to search and seal his papers. Two representatives of the Committee of Public Instruction, Romme and Fourcroy, were present to identify objects deposited with Lavoisier for the purpose of work on the weights and measures and to insure that these were not subjected to sealing. Lavoisier protested with dignity that he had left the Tax-Farm some years previously; that he had declined to receive what was owed to him, and that for three years he had had nothing to do with that organization; that he had since served without salary as Commissioner of the National Treasury, which he had organized; that, if he had given up that office voluntarily, it was to give his time to science in the prosecution of researches on problems of public importance; and that he could not believe that he was among those to whom the decree about sealing of papers was intended to apply, but he was ready to submit to any searching of his papers that was desired, although he wished to state his objections.

The searchers, after completing their task, declared that they had found nothing that could give rise to any suspicion among all the papers written in French; but they had found a packet of letters in other languages, which they had decided to send to the Committee of General Security. Lavoisier insisted on affixing his personal seal to the packet alongside that of the revolutionary representatives, fearing, says Grimaux, that an enemy hand might take the opportunity of inserting some compromising document, but informing the searchers that he took this precaution to insure that the packet was opened only at the Committee of General Security; he stated that he did so, not from distrust, but that all should be in order. This packet was not returned to Lavoisier and it remained in the possession of the Committtee of Public Instruction, to whom it was sent for examination. On September 29 Romme and Fourcroy were directed to examine its contents and a second examination by Fourcroy, d'Aoust, and de Morveau was ordered on November 2. Further, it was not returned with Lavoisier's confiscated possessions after his death; and it seems to have played a part in his condemnation by the Revolutionary Tribunal.

By a new decree passed by the Assembly on September 24, at the instance of the Committee of Finance who were anxious to complete the liquidation of the Tax-Farm, the seals were removed from the papers in Lavoisier's house and in those of the other Farmers-General on September 28, as the Convention had realized that the accounts of the Farm could not be completed unless the Farmers had access to their records. The Farmers, after a loss of four precious months, were now able to resume the work on their papers with some satisfaction, since they were in no doubt about the state of their records and ultimately presented their accounts in good faith and without dissimulation.

But a successful clamor was now raised in the National Convention on September 26 by Antoine Dupin to complete the decree of September 24. He proposed an examination of the accounts of the Farm by a commission of six members, who would offer to denounce the various financial abuses of which he alleged the Farm had been guilty. On the 26th he succeeded in obtaining a vote for additional clauses to the new decree of the 24th. By the terms of these clauses five revising commissioners, who claimed to be in a position to obtain information on the alleged abuses and to furnish proof of the embezzlements that were said to have occurred, were authorized to examine all the accounts for the three previous leases of the Farm, and to submit to the audit office their work on the supposed abuses that they would discover and expose. Those who had denounced the Farm saw in this an opportunity for the State coffers to be replenished from the resources of the former Farmers-General.

Dupin was appointed to the revising commission with one other member of the Convention, who, however, seems to have left the work to Dupin. Dupin's nominees were former employees of the Farm; chief among them was Gaudot, former customs officer at Port-St.-Paul in Paris, convicted in 1789 of embezzling 200,000 livres and of falsifying his records to conceal his thefts. Gaudot had escaped from prison during the riots of August 10, 1792, and was now posing as a patriot thrown into the cells under the monarchy because he alone could reveal how the Farmers-General had robbed the State. Grimaux considered that Gaudot hoped to have access to the documents of the Farm to remove the proofs of his guilt. Gaudot recruited as colleagues, or rather accomplices, some officials of the Farm who had been disappointed in their hopes of promotion or who saw an opportunity of pecuniary reward in joining him. It is most improbable that Gaudot ever knew of Dr. Johnson's aphorism that "patriotism is the last refuge of a scoundrel" and it is certain that Johnson who died in 1784 never heard of Gaudot; but rarely in the record of human iniquity can these words have been more applicable, so rarely that it is almost difficult to believe that the one did not have the other in mind.

The latest decree had fixed the date for the completion of the Farm's accounts as April 1, 1794. The Farmers set to work again on their records, confident that they would be able to disprove any accusations made against them, although all were aware of the hatred in which they were held and some feared false charges that they might not be able to disprove. At the worst, it seemed to them, their wealth might be confiscated. More clamor was now raised in the Convention and, although the latest decree concerning the Farmers' accounts had been passed only two months previously, a new one was passed, again at the instance of an individual without previous reference to the appropriate committee, on November 24, ordering the arrest of the Farmers-General.

Nothing could protect Lavoisier, neither his services to the nation nor his distinction in science. He had been attacked by Marat, and he was suspected as a nobleman, Farmer-General, Commissioner for Gunpowder and Saltpeter, and Academician. Earlier in the year, in March, he had been denounced in the Convention for being at the same time Commissioner for Gunpowder and Farmer-General, but by a saltpeter-maker who probably had other interests than justice on his mind. Later an anonymous informer had accused him of correspondence with an *émigré,* Blizard, but the Revolutionary Committee had satisfied themselves that this was incorrect. On November 4, three weeks before the passing of the decree for the arrest of the Farmers-General, Fourcroy had proposed at the *Lycée* the same kind of "purifying scrutiny" as he had suggested at the Academy of Sciences, and Lavoisier was struck off the list of founders as a "counter-revolutionary" by a committee on which Fourcroy served.

There is some reason to suppose that at this time Lavoisier was informed that the Royal Society of London had considered awarding him the Copley Medal, the highest distinction in the Society's gift. Grimaux said that the award was made in December, 1792, but in this he was mistaken. The Copley Medal was awarded to Rumford in 1792 and to Volta in 1794. There was, significantly, no award in 1793; but the Council would have been considering the award for 1793 in November of that year. A Council Minute of November 21, 1793, records: "The allotment of Sir Godfrey Copley's Medal for the year was taken into consideration, and it was resolved to postpone the giving of it for the present." It was a time, moreover, when the award of a medal from London would have been likely to bring, not distinction, but mortal peril to a Frenchman, especially one so closely associated with the old régime. Council may well have been aware of this and may therefore have hesitated to endanger the life of the great chemist. Grimaux may have obtained his information from Lavoisier's papers, some friend having written to Lavoisier about it, but no further evidence is at present available.

Warrants for the arrest of the Farmers were issued on the day the decree was passed. The accused were ordered to bring with them the papers necessary for the completion of their accounts, but the mass of documents in their office was too enormous. On that day eighteen of the Farmers were arrested and taken to the old convent of Port-Royal, now converted into a prison, Port-Libre, for the detention of suspects. Lavoisier could not be found, since the warrant was made out for his old residence at the Arsenal; he was actually presiding at the Bureau of Consultation. When he was informed that evening of the issue of the warrant, he was on guard duty as a sentry at the Arsenal and, fearing immediate arrest, he took refuge with Lucas, a former usher at the Academy, in the rooms at the Louvre in which the now proscribed Academy had formerly held its meetings. He still hoped that the importance and urgency of his work on the Commission of Weights and Measures would cancel the warrant for his arrest.

Next day, he wrote to the National Convention:

> Representatives of the People.
>
> Lavoisier, of the former Academy of Sciences, left the Ferme Générale about three years ago. Called upon at that time to take up the position of Commissioner of the National Treasury, he contributed largely to the organization of that body. He is now National Commissioner of Weights and Measures.
>
> It is a matter of common knowledge that he never took part in the general affairs of the Ferme which were conducted by a small committee, nominated by the Minister, and, moreover, the works that he has published show that he has always been mainly concerned with the sciences.
>
> He is not one of the Commissioners nominated in pursuance of the decrees for the presentation of the accounts of the Ferme Générale; he cannot therefore be held responsible for the delay with which these Commissioners are reproached; accordingly, he does not think that the decree ordering that the Farmers-General shall be put under arrest until their accounts are presented is applicable to him. In this uncertainty, he begs the National Convention to inform him whether it is intended that he shall concern himself with the accounts of the Ferme Générale, work for which he does not think himself fitted, or whether he is to continue to discharge his duties in the Commission of Weights and Measures, for which he has worked up to now ardently and, he presumes to say, to some purpose.

The letter was not brought before the Convention, but passed to the Committee of Public Instruction, where it was read that evening. Guyton de Morveau presided; but no voice was raised in Lavoisier's favor and the Committee passed to the next item of the agenda without discussion.

Then, on the following day, Lavoisier wrote to the Committee of General Security:

> To the Citizen Representatives of the People constituting the Committee of General Security of the National Convention.
>
> Citizen Representatives,
>
> Lavoisier, of the former Academy of Sciences, is instructed by the decrees of the National Convention to co-operate in the establishment of the new measures adopted by the National Convention.
>
> On the other hand, a newly passed decree orders that the Farmers-General shall be placed under arrest and detained to work on the presentation of their accounts; he is ready to do this, but he considers that he should first inquire which of these decrees he must obey.
>
> The Committee of General Security would ensure the execution of both decrees by ordering that Lavoisier remain under provisional arrest in the custody of two of his brother *sans-culottes*. He points out that three years have passed since he was a Farmer-General and that his character and all his fortune guarantee his moral and physical responsibility.

This was the very committee whose duty it was to insure that the latest decree of the National Convention was carried into effect. Life had flowed past Lavoisier; it seems that he still suffered from the delusion of all reasonable minds, that others are reasonable. No reply, so far as is known, came from the Committee of General Security. On November 28 Lavoisier gave himself up and with his father-in-law, Paulze, joined his former colleagues in Port-Libre. The entry in the prison register for that day reads: "Lavoisier, former Farmer-General. Grounds of arrest: presentation of accounts. By warrant." The reddening tide of revolution was closing on one who was perhaps its most illustrious victim.

In Port-Libre, the older men suffered much from the cold, but the conditions were not severe. Meals were taken in common, and there were sentries at the gates and at the entrances to the corridors, but no bars or bolts, only latches on the doors of the cells. Those detained might meet together in a large room; a roll call in the cells was held at nine o'clock in the evening, but they were not restricted to the cells after that hour. At first the prison was overcrowded, but that condition was remedied, and Lavoisier wrote to Mme. Lavoisier to say that he was making his room more comfortable and that he was preparing his memoirs on chemistry for publication in complete form. "I have begun," he said, "to adopt a way of life fitted to the circumstances in which I am placed; I worked for two hours and a half yesterday afternoon." He advised her not to spend her strength in useless attempts to procure his release, but she continued her efforts without obtaining anything more than permission to visit her husband.

The situation of the Farmers-General was desperate. They could not prepare their accounts without documents and the mass of documents was too great to be brought from the Farm's offices in the Rue de Grenelle to Port-Libre. They appealed to be removed to their offices. On December 18, Borda and Haüy appealed on behalf of the Commission of Weights and Measures for

Lavoisier's release to carry on the work, but the Committee of General Security ignored the appeal. On December 20, the Committee of Public Safety, taking the advice of the Committee of Public Instruction, removed Borda, Lavoisier, Laplace, Coulomb, Brisson, and Delambre, from the Commission of Weights and Measures. The committee on *assignats* and coinage also appealed in vain for Lavoisier's release.

About this time, on December 19, Lavoisier wrote to his wife:

> You give yourself, my dear, much trouble, much fatigue of body and soul, and I cannot share it. Take care that you do not impair your health, for that would be the worst of evils. My life is advanced, I have enjoyed a happy existence as long as I can remember, and you have always contributed to it by the devotion that you have shown me; I will leave after me pleasant recollections. My task is done, but you have a right to hope for a long life, so do not throw it away. I thought yesterday that you were sad. Why be so, since I am resigned to all and look upon all that I shall not lose as gained? Besides, we are not without hope of being together again and, while waiting, your visits give me some happy moments.

As no reply had been made to the petition submitted by the Farmers-General to the Convention on December 2 to be allowed access to their documents in order to complete their accounts, a further petition was submitted. Its discussion on December 11 clearly showed that the Convention still believed that enormous sums could be extracted from the Farmers. The petition, however, was successful, and on December 25 the Farmers were removed to their offices under escort and kept there under guard while they worked at their accounts.

All their property had now been confiscated and seals placed on their houses. Officials visited Lavoisier's house on December 17 and affixed their seals in the presence of Mme. Lavoisier, and then departed for Fréchines for the same duty. The seals affixed at 243 Boulevard de la Madeleine were raised for Guyton de Morveau and Fourcroy, as delegates from the Committee of Public Instruction, to remove in the presence of Lavoisier, brought under guard from Port-Libre, objects entrusted to him for work on the weights and measures. "Who can tell us," asks Grimaux, "what glances passed between the prisoner and the members of the Convention? What feelings must they have had when they saw before them, in the humiliating condition of a criminal, the former director of the Academy, the wealthy Farmer-General, the illustrious scientist whose help and support they had so often asked for?"

At the offices of the Farm, conditions were much worse than in Port-Libre. Some of the prisoners had no beds; most of them now had no money and ate at the charity of their fellows. But at last they had access to their papers and worked for ten hours daily on the preparation of their accounts, on the completion of which most of them supposed that they would be set at liberty. Meanwhile, the revising commissioners, Gaudot and his accomplices, went on with their examination of the accounts and sent brief questions to the accused Farmers, who returned short replies.

After thirty-one days, on January 27, 1794, the completed accounts were sent to the Committee of Finance, while the charges preferred by the revising commissioners were refuted in detail in a lengthy document drawn up by Lavoisier. The Farmers now claimed their liberty, but in vain. Gaudot objected that it was necessary to examine the accounts submitted. After this examination, even Gaudot could not assess the sum alleged to be owed to the State by the Farm at more than 130 million livres, about one-third of that previously suggested by Gaudot himself and noised in the rumors of recent years. Other charges, however, were now preferred, namely, excessive rates of interest, delays in the payment of money to the Treasury, and excessive adulteration of the tobacco with water.

Details of these three new charges demand examination. The first alleged that, during the lease of David (1774-80), the Farmers allowed themselves interest at a rate of 10 per cent, while the terms of the lease gave them only 4 per cent. This charge was a deliberate confusion. When they entered on this lease, the financiers had to pay into the Treasury a sum of 93,600,000 livres, of which 72,000,000 was an advance to the Government and the remaining 21,600,000 was due to the holders of the previous lease as purchase money paid by the holders of the new lease for the value of the plant and other property at the salt works and tobacco factories. Every Farmer-General, and there were then sixty, had to provide 1,550,000 livres. As few were able to furnish this amount, most of them had recourse to lenders who were only too ready to put their money into such safe hands. When the terms of the lease were drawn up in conference with the Controller-General, the Farmers included in their expenses the interest that they had to pay on their loans. For this total sum of 93,600,000 livres, they set down as expenses the interest on 60 millions at 10 per cent and on 33,600,000 at 6 per cent. This was included in the terms of the lease as expenses, but the Government was not bound to accept the terms proposed by the Farmers, since the lease was freely negotiated after discussion. Moreover, this practice was normal and legal and had been followed up to the time when Necker changed the method of paying the Farmers; the interest was paid, not by the Treasury, but out of the Farmers' profits. Interest at 4 per cent, which the terms of David's lease applied to the 72 million livres paid as an advance into the Treasury, was something that the Farmers never received, for the Government added the charge to the price paid for the lease. Two different clauses had been wilfully confused.

As for the alleged delay in the payment to the Treasury of money received by the Farm, with the suggestion that it was meanwhile used in private speculation, the evidence offered for this was the dating of the receipts. As the dates on these documents very properly bore the final date of clearance after all accounts had been examined and verified, and not the actual date of payment, this allegation was easily rebutted. Besides, the Farm was always in advance in its payments to the Treasury.

There were other minor and similar allegations of the same nature, all refuted in detail by documentary evidence; but the third important charge played a great part, or seemed to do so, in the condemnation of the Farmers-General. It alleged the excessive adulteration of tobacco by increasing, beyond the legal proportion allowed, the amount of water, the *mouillade* or moisture, added to the tobacco after grating. (Raw tobacco, as received by the manufacturer, is sun-dried and matured and contains about 11 to 12 per cent of water; it is too dry to handle and water must be added, but much of this added water is removed in the process of manufacture.) This practice, it charged, increased the price of tobacco by making the citizen pay for the added water as tobacco, "an action as perilous to the health of the consumer as it was damaging to his interests." This was an old sore; and, in one sense, the *mouillade* can be said to have brought death to the Farmers-General and the title of Dupin-Mouillade to their enemy.

Grating of tobacco by the Farm had been passionately disputed for many years; and France was divided into "graters" and "anti-graters." All tobacco had formerly been sold in plug, but when snuffboxes became fashionable a demand for powdered tobacco arose. A monopoly for the sale of tobacco had been granted to the Farm. (The system for tobacco sale which still exists in France is a survival of this practice.) Much contraband tobacco was sold and grated in the retailers' shops, however, and the Farm revenues steadily decreased, despite increased consumption. When Farmer-General Delahante proposed that grating be forbidden to retailers and restricted to the Farm, his proposal was accepted by his colleagues with some doubt of its expediency. Public opposition was loud, because the retailers had established grating rooms at their shops and gave employment in this work, and it was even suggested that tobacco grated in factories, such as those of the Farm, was injurious to health. So, in one way, the charge was an old one in a new form; but it already had a long history.

In 1774 the chemists Baumé and Cadet, on the direction of the Minister, had examined the tobacco sold in the depôt at Morlaix and found that it was inferior. In 1778 Lavoisier had drawn the attention of his colleagues to the unsatisfactory state of affairs in this department of their administration, reporting: "I am afraid that the *mouillade* exceeds what is proper. . . . one must be extremely careful with it and it is better to make small profits than to run the risk of causing public dissatisfaction." Before 1786, the net amount of water added was one-seventeenth of the tobacco sold to the retailers; the retailers were charged for only sixteen of every seventeen ounces that they bought from the Farm. After 1786 the proportion of water added was lowered to about one-fortieth. (Modern pipe-tobacco contains from 6 to 12 per cent, or in some kinds more, of added water.)

As a further means of reducing contraband, the Farm endeavored to drive out bad tobacco with good. They removed all stems and damaged leaves from the raw material, thereby improving the quality considerably, since 100 pounds of leaves now gave only 75 to 78 pounds of grated tobacco. As Lavoisier pointed out, this very practice refuted the charge of excessive *mouillade,* since the Farm would scarcely go to the trouble of rejecting such a high proportion of raw material, a third of the finished product, with the aim of improving the quality of the tobacco, and then add water beyond the proportion demanded by good manufacture.

The revising commissioners, however, claimed that the Farm had unfairly acquired 30 million livres by excessive *mouillade,* which they estimated at 14 pounds in 100 pounds of grated tobacco, disregarding losses by evaporation during manufacture and the practice of selling to retailers on the calculated dry weight. Lavoisier easily dealt with this charge. He showed correctly that most of the water added in the grating had evaporated before sale in the shops, and that there remained at first only 6 per cent and later 2 per cent, and he pointed out that the Farm sold its tobacco to the retailers at a price calculated on its dry weight.

There was still another charge. At the time of the suppression of the Farm in 1791, the National Assembly had allocated 48 million livres in payment for the salt works and tobacco factories that were taken over by the nation, but, after the Farmers-General had later found that their value was only 26 millions, the Farmers, on the advice of Lavoisier, returned the sum of 22 millions to the State. Now, however, the revising commissioners boldly stated, without any proof, that this 22 millions was "a trust for the benefit of the tyrant" and that surrender of this sum had been agreed to only after their own researches. Words mean nothing to those who make charges of this kind or perhaps they have unusual meanings. But the sum total, according to Gaudot's studies and allegations, of moneys owed by the Farmers-General to the nation and amassed by alleged theft amounted to 130 million livres. This was the subject of the ensuing discussion in the Convention.

The report went first to the Committee of Finance which studied it for two meetings. The chief of the claims office of the Tax-Farm, devoted to his old employers, twice presented himself before the Committee and offered to give all necessary explanations on behalf of the prisoners. He was refused a hearing and ordered out at once, as the proceedings were secret. Dupin, questioned by the wives of some of the prisoners, replied that all would be set at liberty soon, although he was at this very time making preparations to get them prosecuted before the Revolutionary Tribunal.

France was now ruled by the Terror. A new law had declared suspect all persons who were of noble birth, or who had held office before the Revolution, or who had any connection with an *émigré,* or who could not produce a certificate of *civisme,* of support of the Revolution, from the local authority, who, of course, granted this or refused it at discretion. The Revolutionary Tribunal was used as an instrument of the Convention to get rid of every doubtful element; trials were a farce and the chance of acquittal became ever more faint. Philippe, Duke of Orléans, "Égalité Philippe," had been guillotined, and Madame Roland, and, with almost incredible cruelty, Bailly, first National President, first Mayor of Paris, astronomer, and Academician. The Commune of Paris was striving for supremacy with the Committee of Public Safety, whose most active member was Robespierre. Fouquier-Tinville, President of the Revolutionary Tribunal, sent even the founders of the Republic, Hérault de Séchelles, Danton, Desmoulins, and Westermann, to execution on charges of complicity with the foreigner.

Yet the Farmers-General, although they were well aware of the hatred fostered against them, were not without hope. They had committed no act of *incivisme* and they had not been involved in politics or with any party. Lavoisier prepared his defense. At the same time he asked for a certificate of his services from the Bureau of Consultation. The Bureau appointed a committee, consisting of Coulomb, Servières, and Hallé, to draw up a report. It was adopted at the meeting on April 23, presided over by Lagrange; those present included Borda, Desaudray, Leroy, Silvestre, Trouville, Jumelin, and Dumas. It was resolved to honor Lavoisier by entering this report in the records and sending him a copy. The report read:

> The Bureau of Consultation of Arts and Crafts, after having heard the report of its committee on the request and on the works of Citizen Lavoisier, considering the number and the importance of the discoveries of this citizen, the great and useful revolution that they have contributed to bring about in chemistry, the light that they have shed on the nature of many substances not well known until our time and on the chief phenomena of vegetation and of animal economy, the advantages that have thereby resulted in

almost all the useful arts that have any connection with chemistry, such as painting, assaying, and the working of mines, etc., and finally, that the opinion of most of the scientists of Europe assigns to Citizen Lavoisier a distinguished place among those who have brought honor to France, and considering also that Citizen Lavoisier has shared zealously and assiduously in the labors of the Bureau of Consultation to insure to craftsmen the rewards due to their talents, has resolved that this testimony of the esteem in which he is held shall be recorded in its minutes and that a copy shall be sent to Citizen Lavoisier.

Lavoisier applied also to the national agents for gunpowder and saltpeter, Le Faucheux, his former colleague in the now defunct Commission, and Champy. Both of them tried to save him from the Tribunal on the grounds that as a former Commissioner of Gunpowder he was among those designated by a resolution of the Committee of Public Safety as required for the completion of the Commission's accounts.

To two of his old colleagues in the Academy of Sciences, Cadet and Baumé, the latter a passionate opponent of the new chemical theory, Lavoisier also appealed. They came to him in prison with a document certifying that he had always been opposed to the *mouillade* in the grating of tobacco, and that they were aware of this through the official inquiry that they had made in 1774.

There were others in higher places, old friends and associates, who soon showed themselves unwilling to take any action to protect or to save him, while these mentioned here did so in equal peril of their lives.

Lavoisier himself drew up a short account of his career. He summarized his work during the Revolution, his chemical and agricultural researches, the part that he had taken in the Provincial Assembly of Orléanais, in the Constituent Assembly, in the Commune of 1789, in the National Treasury, at the Bureau of Consultation, in the Committee on *Assignats* and Coinage, on the Commission of Weights and Measures, on the Committee on Public Health, together with his services in the National Guard at the beginning of the Revolution.

Meanwhile, the time for Dupin to present his report to the National Convention was drawing nearer. Some of Lavoisier's friends wished to make representations on his behalf; but he declined because he feared that they might thereby compromise themselves. Loysel, a member of the Convention and former director of the factory at St.-Gobain, made a vain attempt to save one with whom he had been friendly. But a far bolder scheme was devised by his old friend, Pluvinet, to whose shop in the Rue des Lombards Lavoisier had often come to buy materials for his laboratory. Pluvinet's idea was to approach Dupin himself and his plan almost succeeded. He was acquainted with Dupin's sister-in-law, a lady of unconventional manners, who agreed to play her part, which she did very successfully. She obtained from Dupin a promise that Lavoisier would be separated from his colleagues and transferred to another prison, and that the report would not be in any way unfavorable to him. Dupin, however, complained that Mme. Lavoisier had not approached him personally, but had canvassed his help only through aristocrats like herself and had never condescended to call on him. Warned by Pluvinet, and in defiance of the decree forbidding all ex-nobles from entering Paris, Mme. Lavoisier called on Dupin. Instead of pleading, however, she declared that she had not come to beg for mercy for her husband, that he was innocent, and that only scoundrels could accuse him. "Lavoisier," she said, "would be dishonored if he separated his case from that of his colleagues; there are designs on the lives of the Farmers-General in order to get hold of their fortunes; if they perish, they will all die innocent men." Dupin, irritated by this attitude, was deaf to all further pleas. In preparing his defense, however, Lavoisier had in some degree already separated his case from that of his colleagues. Mme. Lavoisier has been blamed for her bearing at this interview, but it must be remembered that her father's life as well as her husband's was at stake; and Grimaux rightly recommended the critic to turn his thoughts to the prudent silence of Lavoisier's former associates.

On Monday, May 5, 1794, Dupin submitted to the Convention a long indictment of the Farmers-General. Based on the report of the revising commissioners, it repeated the charges but made scant mention of the rebuttals. Although Dupin indicated that some of the Farmers were opposed to the excessive addition of water to the tobacco and that the Tribunal would be able to distinguish between them, he appeared to regard them all as guilty. He blamed them also for the order in council of 1774, even though it was, in fact, imposed by Terray after the Farmers had signed the new lease. It will be recalled that they had accepted it under the threat that their advances would be returned only in the form of life annuities. Ignoring the difficulties under which they had labored to present their final accounts, he suggested that, in disobedience to the Convention, they had put off the work for two years because they were eagerly awaiting the return of the old régime. Speaking on behalf of the committees concerned and the revising commissioners, and skillfully manipulating the figures, he dragged the Convention along with him. Without discussion they decreed that the Farmers-General should be arraigned before the Revolutionary Tribunal.

That tribunal was, however, it may be noted, competent to deal only with acts of *incivisme,* with which the Farmers had not been charged, not even by Dupin.

It was after four o'clock when the decree was passed. An observer hurried to the office of the Farmers-General; the first whom he met was Lavoisier, to whom there then fell the unhappy duty of informing his colleagues of what they had at last foreseen and now daily expected. They burned their personal papers at once. Dupin gave instructions that no one was to be allowed to see them, adding that the decree could not be carried out for three or four days. It was, in fact, not presented to the Revolutionary Tribunal until the next morning, May 6, and not registered until May 7. Legal delays were, however, not to be allowed to stand in the way of the execution of the Farmers.

Irregularities in the arrests had already occurred. One member of the Farm was released after arrest on the grounds that he was an assistant, not a Farmer-General; three others were arrested because they were assistants, but later released. Some of the warrants were issued by the Commune of Paris, others by the Committee of General Security. The latter body ordered the arrest of Baudon—who had been dead for fourteen years. One man arrested as a Farmer-General had never had any connection with the Farm. Some were arrested months later; fourteen were never arrested, either because they were not sought for or because they had evaded pursuit. The warrant for Lavoisier's arrest was made out for him as residing at the Arsenal.

But these irregularities were trifles in comparison with what now took place: Fouquier-Tinville, President of the Revolutionary Tribunal, signed the indictment on May 5, the very day on which the decree was passed and ordered the immediate removal of the prisoners to the Conciergerie. Later this was one of the main charges against him: the decree was not registered until May 7, but he signed the indictment and acted in accordance with the decree two days previously. He said that he had called on the Committee of Public Safety at two o'clock in the night of May 5-6, but this defense was inadequate, since he had quoted the very terms of the decree in the indictment that he signed on May 5. To this criticism, he replied: "You want to bring me to account, because I brought those blood-suckers of the people and counter-revolutionaries to their account." Grimaux concluded that the irregularity was even worse than this, claiming that Fouquier's indictment was already prepared before the Convention had passed the decree. He based this conclusion on the facts that Dupin's report was read at half-past three o'clock and that the prisoners were already being moved to the Conciergerie at seven o'clock, and declared that it was improbable that three hours, apparently allowing half an hour for Dupin's report, would give enough time for the pronouncement of the decree to be conveyed to Fouquier and for the latter to draw up the lengthy indictment, give orders for the removal of the prisoners, and secure instant obedience to his commands. In support of this conclusion, he pointed out that the indictment was couched in the terms of Dupin's report, whence it followed that Dupin had communicated it to Fouquier before he had read it to the Convention, so that it was really Dupin who "relentlessly sought the death of the Farmers-General and guided the pen of the Public Prosecutor." It is difficult to disagree with this conclusion.

Some of the Farmers-General thought to avoid the torture that they knew to be before them by taking their own lives with opium. When they suggested to Lavoisier that he share their fate, he persuaded them against suicide. "Why go to meet death?" he said. "Because it is dishonorable to receive it by the decree of another, especially by an unjust decree? For us, the very outrageousness of the injustice wipes out the dishonor. We can all face it with confidence in our private lives and in the judgment that will be passed on them, perhaps in a few months; our judges are neither in the tribunal that summons us nor in the rabble that will insult us; a pestilence is ravaging France; at any rate it strikes its victims down at a single blow; it is about to attack us, but it is not impossible that it may stop before some of us at least. To take our own lives would be to acquit the madmen who are sending us to death. Let us think of those who have preceded us to the scaffold and at all events leave a good example to those who will follow."

Mollien, an official of the former government attached from the Controller-General's office, witnessed this scene. He was detained in the Farmers' offices and expected to share their fate; in fact, although not arraigned, he had been one of those who had contemplated suicide. But he survived, owing his life to Lavoisier's dissuasions.

Soon the mounted police arrived, followed in half an hour by four large covered wagons. The roll was called by the concierge and, in batches of four, as they replied to their names, the prisoners were shut into the wagons, the municipal officials in charge of the transfer of the prisoners drinking and laughing meanwhile in the concierge's room. More than an hour was taken in completing these formalities and night had fallen when the procession, escorted by a double file of mounted police, started on its way, lit by men on foot carrying torches. The next name on the prison calendar was Mollien, who expected to be removed with the others; but when the roll call for the Farmers-General had ended, the warder Nécard pushed him back, muttering "You have nothing to do with this." Nécard also took care not to mention Mollien to the Committee of General Security. "One must console oneself," he said, "by one good action for so many others."

The wagons rumbled through the night along streets familiar to Lavoisier since his boyhood, half-way across the Pont-Neuf, to the Conciergerie on the Île de la Cité, which they reached at eleven o'clock. After more long formalities, some of the prisoners were put into cells and others into the room in which Marie Antoinette had been imprisoned before her execution. Some had bare trestle beds; most slept on benches or on the floor. All were benumbed, as the weather was bitterly cold. At seven o'clock on the morning of May 6 the cell doors opened and they were allowed to assemble in Marie Antoinette's room; towards afternoon they set up two tables there and made themselves less uncomfortable. Fortunately, also, through Dobsen, a judge of the Revolutionary Tribunal and a relative of the prisoner Delahante, they were able to get three rooms for the following night. These were shared among seventeen prisoners, each of whom was provided with a bed, a mattress, and a quilt; six pairs of sheets which Delahante had secured he distributed among his friends and the oldest of the prisoners. The smallest room was allotted to Paulze and Lavoisier.

Next morning, May 7, at half-past seven o'clock, they were all taken to the Registry and thoroughly searched to insure that they carried no weapons, then led into a room adjoining the Tribunal and separately interrogated. Lavoisier appeared before Dobsen in the presence of the Registrar, Ménot, and the Public Prosecutor, Fouquier. As Grimaux wrote, the very brevity of the interrogation shows that it was a mere formality to satisfy the scruples of the law and identify the accused. The records read:

> Also appeared Antoine Laurent Lavoisier, aged fifty, born in Paris, former Farmer-General and member of the former Academy of Sciences, living in Paris, Boulevard de la Madeleine, Section des Piques.
>
> Asked, What department he had charge of?
>
> He answered that he was head only of the departments of Lorraine, Les Évêchés, and the Domaine de Flandre.
>
> If he had not been guilty of peculation of Government funds, of extortion, of adulteration, and of fraud against the people?
>
> Replied that, when abuses had come to his knowledge, he had reported them to the Minister of Finance, particularly abuses relating to tobacco, which he was in a position to prove from authentic documents.
>
> If he had chosen a counsel for his defense?
>
> Replied that he knew none of them, and we appointed Citizen Sézilles for him.
>
> This being read over, he signed with us and the registrar.
>
> Dobsen. A. Fouquier. Lavoisier. Ménot.

One stage in the farce was over. A month later, to speed the course of justice, the worst law of the Terror was passed; to avoid the paralysis of the Revolutionary Tribunal by such formalities and such delays, it abolished the defense of prisoners by counsel and the examination of witnesses. Judges and jury were empowered to decide by the impressions that accused persons made on them, and

29. Lavoisier in 1793 from Mlle. Brossard Beaulieu's engraving of a sketch said to have been made during his detention: below is his autograph from a document of 1771.

the penalty for all offences was increased to death. When the Farmers-General were arraigned, however, matters had not become so simple or so swift; and some formalities were acknowledged, even if only in the letter.

On their return to the Conciergerie, the Farmers-General, now deprived of their *assignats,* asked for the bread supplied to prisoners, but some friend had arranged for a good meal to be sent to them. Although they had not been told when they were to appear before the Tribunal, all thought that their last night had come. It was probably at this time that Lavoisier wrote to his cousin, Augez de Villers, the last of his letters of which we have any record:

> I have had a fairly long life, above all a very happy one, and I think that I shall be remembered with some regrets and perhaps leave some reputation behind me. What more could I ask? The events in which I am involved will probably save me from the troubles of old age. I shall die in full possession of my faculties, and that is another advantage that I should count among those that I have enjoyed. If I have any distressing thoughts, it is of not having done more for my family; to be unable to give either to them or to you any token of my affection and my gratitude is to be poor indeed.
>
> So it is true that the practice of every social virtue, important services for one's country, a life spent advantageously in the advancement of the useful arts and of human knowledge, are not enough to protect a man from a sinister end or to avoid dying like a criminal!
>
> I am writing to you today, because tomorrow perhaps I may no longer be allowed to do so, and because it is a comfort to me in these last moments to think of you and of those who are dear to me. Do not forget

that this letter is for all those who are concerned about me. It is probably the last that I shall write to you.

It was at this time, too, that a deputation from the *Lycée des Arts* visited him in prison to confer on him a now unidentifiable distinction awarded to him three days previously. This display of courage on behalf of the most illustrious of their members is refreshing amidst the miserable knavery that surrounded Lavoisier in his last hours.

At one o'clock next morning, May 8, the thirty-two prisoners, after a roll call, were each given a copy of the indictment laid against them. The writing was small and scarcely legible and, in any case, they were ordered to put their lights out. At dawn, they were taken to the Registry and thoroughly searched, and their last personal possessions were taken from them; they were prepared for the scaffold before they were tried. Thirty-one of them passed on to a room adjoining the Tribunal, Farmer-General Verdun's name having been removed from the list by order of Robespierre, his protector. There they were introduced to the four official counsel for the defense; one quarter of an hour was allowed for consultations with the accused and preparation of their defense.

At ten o'clock they were brought before the Revolutionary Tribunal, with Coffinhal presiding. There was the customary display of National Guards with fixed bayonets and the usual interested crowd. Coffinhal was assisted by two judges, each at a separate table on which was a bottle and a glass. Gilbert Liendon acted as Public Prosecutor. Lavoisier's appointed counsel did not appear, but another had taken his place among the four already mentioned: it was a mere formality in any event. Coffinhal asked some brief questions resembling those of the previous day and then went on to ask each one whether he was a noble and what he had done since the Revolution. Replies were treated with derisive comments and false interpretations by both the jury and the judges. This phase lasted about an hour and a half and the session was suspended for twenty minutes.

On its resumption, the indictment was read and Liendon asked a question that the prisoners did not understand. Coffinhal called on one of them, Sanlot, who replied that he was only an assistant in the Farm, that he had left it more than ten years ago, and that he was not in a position to make a reply. Another, Delaage, referred the President to the reply already made by the Farmers in their rebuttal of the charges. "Let us see," said Coffinhal, "if M. de St.-Amand, who ruled so despotically over the Tax-Farm, is better able to reply." St.-Amand replied that he did not understand the question, and then Liendon declared that he would abandon it. He proceeded to reproach the Farmers-General with having presented to the Minister false statements on the proceeds of an old lease as a basis for the price of a new one which was to be negotiated. St.-Amand readily replied that the prices of leases were based on returns made in the Ministry of Finance and not on the Farmers' own figures. Coffinhal interrupted loudly and told the prisoners that they were to answer "Yes" or "No" and nothing else, that the case would proceed without intermission, and that they would not be permitted to go into details to gain time for themselves. All individual or collective defense was forbidden.

The prisoners realized that not one of them would escape death, except perhaps Didelot, who thought to the very last moment that he would be set free. He had been put in charge of the liquidation of the finance company that was concerned with Crown lands and, during the imprisonment of the Farmers-General in their office, he had been taken to his work every day under escort. In fact, he had not been removed to the Conciergerie with the others. After naively consulting Dupin, who assured him that he would be acquitted, he quite confidently gave himself up at the Conciergerie, although he had not belonged to the Farm for fourteen years.

Just as Coffinhal had interrupted St.-Amand, Liendon rose to read a decree that had just been passed by the Convention, ordering the release of three of the accused because they had not been Farmers-General, but assistants, and as such had not signed any of the leases on which the charges were based. The decree had been obtained through Dobsen who, after vainly pleading with Fouquier to remove the names of these three from the list, had hurried to Dupin's house and dragged him to the Committee of General Security, which drew up its report at once. Dupin had then presented it to the Convention and secured its immediate passage without debate and its transmission to the Tribunal, "the only honorable action that can be put to his credit," said Grimaux. The father of one of those set free, de Bellefaye, was a Farmer-General; his last happiness was the sight of his son escaping death.

This incident over, Liendon asked a few further trifling questions and then proceeded with his speech for the prosecution in the ranting bombastic rhetoric of the day. He recalled the various charges of extortion and adulteration, and concluded by declaring that "the record of the crimes of these vampires is complete; their crimes clamor for vengeance; the immortality of these creatures is burned into public memory; they are the cause of all the evils that have for some time afflicted France." Then the four counsel for the defense were heard—for twenty-eight prisoners *en masse* after a consultation of fifteen minutes in a case of which they had no previous knowledge! Legal formalities, however, must still be observed. Hallé submitted to the Tribunal the report of the Bureau of Consultation, but it was apparently considered beneath notice, unless it is true that Coffinhal disposed of it, as seems probable, with the words since attributed to him: "The Republic has no need of men of science."

The Revolutionary Tribunal, being a special criminal court, had no powers to deal with any of the offenses with which the Farmers-General were charged, but only with counter-revolutionary activity. The cunning distorted genius of Coffinhal made up for that, however. Addressing the jury, he asked: "Has there been a conspiracy against the people of France tending to favor their enemies by all possible means, by practising all kinds of extortions and adulterations on them, by mixing with the tobacco water and ingredients injurious to the health of the citizens, by taking 6 and 10 per cent, both for the interest on the different guarantees and on the capital necessary for the working of the Tax-Farm, while the law allows only 4 per cent, by withholding sums that ought to have been paid into the national treasury, and by plundering and robbing by every possible means both the people and the national treasury with the object of despoiling the nation of the immense sums necessary for the war against the despots raised against the Republic and of handing these sums over to them?"

Coffinhal had thus, as judge, invented a new charge, not mentioned in either Dupin's report or the indictment —counter-revolutionary activity in complicity with the foreigner. His twisted genius had an imaginative side, for counter-revolutionary activity was punishable by death, whereas peculation and such crimes were neither capital offenses nor within the competence of his tribunal. His addition to the indictment on tobacco—of "ingredients injurious to the health of the citizens"—was relatively a trivial flourish.

The jury unanimously found the accused guilty. Time pressed. The tumbrels waited for the day's batch of victims to be taken to the Place de la Révolution. The judges were in such a hurry that the verdict was not entered on the record of the judgment, as Dobsen showed later at the trial of Fouquier-Tinville.

Coffinhal pronounced sentence:

> The verdict of the jury is that it is proved that there has been a conspiracy against the people of France tending to favor, by all possible means, the success of the enemies of France; . . .
>
> That Clement Delaage, Danger-Bagneux, Paulze, Lavoisier . . . [here follow the names of the other Farmers-General to the number of twenty-four] . . . are all convicted of being principals or accessories in this conspiracy;
>
> The Tribunal, having heard the Public Prosecutor on the application of the law, condemns the aforementioned to the penalty of death, in pursuance of article 4 of the first paragraph of section 1 of the fifth part of the penal code, which he has read and which is worded thus: *All intrigue, all correspondance with the enemies of France tending, either to promote their entry into the Dependencies of the French Empire, or to surrender to them towns, fortresses, harbors, ships, magazines or arsenals belonging to France, or to supply them with help in soldiers, money, rations or munitions, or to favor in any manner the advance of their arms on French territory or against the land or sea forces, or to undermine the allegiance of officers, sol-*

diers or other citizens towards the French nation, shall be punished by death,

Declares the property of the condemned confiscated to the Republic. . . .

Orders that at the instance of the Public Prosecutor the sentence shall be carried out within twenty-four hours. . . .

The last touch to this legal farce was the reading of this irrelevant section of the code. This was the article by which the Revolutionary Tribunal struck down its victims; it had been invoked to send the Dantonists to the scaffold.

Lavoisier and his colleagues were therefore not condemned for any actions or alleged misdeeds as Farmers-General. They were condemned in a mass trial during the Reign of Terror, with token observance of legal forms of prosecution and defense, on a charge of counter-revolutionary conspiracy and supplying the enemies of the Republic with immense sums allegedly withheld from the National Treasury. There was, of course, no evidence for this; but it was a time when every man suspected his neighbor and when the very founders of the Republic, even Danton himself, had been sent to the guillotine on charges of complicity with the foreigner.

Judgment having been pronounced, the victims were taken back to the Conciergerie and conducted into the tumbrels, which were driven over the Pont-Neuf, past the Louvre, meeting place of the former Academy of Sciences, and past scenes known to Lavoisier since his earliest days; across the river was the Mazarin College, his old school; and so on to the Place de la Révolution. All were silent, except Papillon d'Auteroche, who seeing the crowd in the Revolutionary dance and song, the *carmagnole,* said: "What vexes me is to have such unpleasant heirs." The crowd is said to have made no demonstration as the procession passed and the bystanders were moved to pity rather than insult.

The Farmers-General were executed in the order in which their names appeared in the indictment—Paulze third, Lavoisier fourth. While awaiting his own fate, Lavoisier saw his old friend and colleague, his father-in-law, executed before his eyes. Bodies of the victims were thrown into nameless graves in the cemetery of Parc Monceaux.

Thus perished Lavoisier on May 8, 1794, one of the greatest sons of France. "Only a moment to cut off that head," said Lagrange to Delambre next day, "and a hundred years may not give us another like it."

XXVII

EPILOGUE

CRITICISMS AND APPRECIATIONS LIKE THAT OF LAGRANGE had to be whispered while the Terror raged through France. The journals indulged in insults; and, according to Barante, the *Orateur du Peuple* compared the blood that streamed from the scaffold "with the beds of purple on which the Farmers-General had reclined in ease and luxury." Only a few days after Lavoisier's execution, however, the *Décade Philosophique,* in reviewing the newly-published second edition of his *Directions for the Manufacture of Saltpeter,* praised the work highly as "a treatise written in a pure, clear, concise style; . . . very useful in spreading the real principles of the art of the saltpetermaker," and added, *"Bread, iron, and saltpeter* are all that Republicans need," but it did not dare to mention the author's name, although at that very time the Government was utilizing in national defense the rapid processes which he had devised for the refinement of saltpeter.

Nearly a century later, Grimaux made a long study of many documents and of almost every detail of the last five months of Lavoisier's life, from his arrest to his execution. Deeply moved at the culpable waste and untimely loss of a great mind at the height of its powers, he asked himself whether highly-placed friends could not have saved Lavoisier and concluded that history was right in reproaching for their inactivity the scientists who had been associated with him and who knew the powers of his mighty mind and the fineness of his character. Mme. Lavoisier openly accused these old colleagues of being responsible for her husband's death. Lalande wrote enigmatically: "His prestige, his reputation, his place in the Treasury set him above other men, but he used his position only for doing good, though it did not prevent him from making many enemies. I would like to think that they did not play their part in his destruction." Lavoisier was not suddenly struck down without warning. During five months not one of his old associates then in power intervened on his behalf—neither Monge, who was closely in touch with Robespierre; nor Hassenfratz, who was a leading member of the Jacobins; nor Guyton de Morveau, who was a deputy in the National Convention; nor Fourcroy, who also sat in the Convention.

Fourcroy was, indeed, not long afterwards accused of demanding Lavoisier's death, but the pamphlet in which this charge was made was written by Sacombe, who had a grudge against Fourcroy and who was later convicted of libelling Baudelocque. Grimaux himself stated that there was no evidence for this unjust accusation, although he found much to condemn in Fourcroy's negligence. Fourcroy's response to this attack was dignified; and Chevreul, speaking from an acquaintance of seven years, remarked that he had nothing against him and that his friends remained faithful to him, while Thibaudeau spoke well of his moral character.

Grimaux, however, felt that history could not judge him so favorably. Writing of Fourcroy's character, he maintained: that he was vain and ambitious; that he rode to eminence on the exploitation of Lavoisier's new system of chemistry by his quick intelligence and his facility in writing; that he was a second-rate chemist but an unrivalled teacher; that it would be unjust not to recognize his great gifts in exposition, but that the reputation that he thus acquired was a mere reflection of the glory of his master; and that historians had confounded the skillful popularizer of the new chemistry with the genius who had created it. This verdict seems a just one; we have already seen that Lavoisier had had to express his dissatisfaction with Fourcroy's description of the new theory as "the theory of the French chemists."

Fourcroy's conduct in the discharge of his public and official appointments during the last period of Lavoisier's life reveals the political trend of his mind. His proposed membership purge of the Academy of Sciences in 1792 had revealed his real thoughts on such matters. In 1793 as a member of the National Convention and the Committee of Public Instruction, he had helped to destroy the Academy; and he had continued with his "purifying scrutinies" at the Society of Medicine and the *Lycée*. As his political career advanced, he became more and more, either from ambition or from fear, the loyal supporter of the Revolution, although it is to be doubted whether he had any deep political convictions. His make-up might well have contained something of the fawning courtier; indeed, his death was said to have been caused, in part, by his excessive grief over his supposing that he had incurred the displeasure of Napoleon.

We have seen in previous chapters how often Fourcroy appeared in authority in public matters where Lavoisier was concerned, and how invariably he took the opposite course or, more strictly speaking, allowed matters to take the opposite course. Grimaux, however, anxious to be just to a fellow countryman, recognized that Fourcroy was not lacking in private virtues, that he had saved the chemist d'Arcet, and that he did memorable work on the Committee of Public Instruction. Perhaps his most incomprehensible action was his funeral oration on Lavoisier at the *Lycée des Arts* on August 2, 1796, after the tide had turned, when he said: "Let us carry ourselves back to those frightful times. . . . when the Terror separated friend from friend and isolated members of families even round their own hearths, when the slightest word, the least feeling of sympathy for the unfortunates who preceded us along the road to death, was a crime and a conspiracy." Grimaux considered that this revealed the true weakness in Fourcroy's character—fear. He concluded that Chevreul's opinion that no action could have saved Lavoisier was unacceptable.

Certainly, he agreed, action would have been too late at the eleventh hour. But if those members of the Convention who had been friends or followers of Lavoisier had combined in representations to Robespierre, to the Committee of Public Safety, to the Committee of General Security, or to Dupin; if they had recalled the great discoveries of Lavoisier, his services to France, so evident by the progress that he had brought to pass in the production of saltpeter and the manufacture of gunpowder; if they had boldly declared that it was urgent to requisition his labors for the service of the Republic, who could say that their views would not have been listened to? And he went on to point out that Borda, suspect as an ex-noble, and Haüy, a former priest, had protested against Lavoisier's arrest; that Hallé with other members of the Bureau of Consultation had appeared in Lavoisier's favor even before the Revolutionary Tribunal, while Monge, Hassenfratz, de Morveau, and Fourcroy had remained silent; and that Pluvinet, an obscure man, had succeeded in his plan of extracting a promise from Dupin to save Lavoisier from the scaffold. Would Dupin, asked Grimaux, have therefore been deaf if he had been approached by his fellow representatives in the Convention? Could Lavoisier not have been saved when only a word from Robespierre caused Fouquier-Tinville to remove Farmer-General Verdun's name from the indictment? To these strictures there seems no answer. Lavoisier's fellow scientists in the Convention were morally responsible for his death.

Since Grimaux wrote, however, some other evidence has come to light. When representatives of the local section of the Revolutionary Committee made a domiciliary visit to Lavoisier's house on September 10 and 11, 1793, to search and seal his papers, Grimaux records that they removed a packet of letters "written in English," which was sent by the Committee of General Security to the Committee of Public Instruction for examination. The latter committee on September 29 ordered Romme and Fourcroy to examine these letters, and on November 2 ordered a further examination by Fourcroy, d'Aoust, and Guyton de Morveau. Grimaux located this packet in the Archives Nationales and noted that it included also a letter in Italian from Spallanzani. The author has been able to study these letters recently by the courtesy of the Director of the Archives and has found that they contain much more than Grimaux indicated and possibly material that may have led to Lavoisier's death.

The seventeen documents in the packet included the

following letters: one from Franklin to Mme. Lavoisier; two from Joseph Black to Lavoisier; one from Robert Kerr to Lavoisier; two from Josiah Wedgwood, the potter, to Lavoisier, with a French version of one of these in Mme. Lavoisier's hand; two in Italian—one from Agostino Vivorio, Secretary of the Società Italiana, to Lavoisier to announce his election to the Society and the other from Spallanzani, a prominent Italian scientist, with a French version; one from Dr. Gillan of London to Mme. Lavoisier; one from Joseph Priestley to Lavoisier; one from Walter Berry of London to Lavoisier, enclosing a letter of Berry's to Swediaur concerning the publication of a medical book; an unsigned letter in French to Mme. Lavoisier; and another in French from Méchain to Lavoisier, enclosing a document in Spanish, authorizing Méchain to continue his work on the Spanish frontier. The contents of most of these letters could scarcely arouse any suspicion in normal times, but the times were far from normal. Every man suspected his neighbor. The constant dread was counter-revolutionary activity in complicity with the foreigner, the existence of enemies within the gate in the form of what has since come to be called "the fifth column."

The letter from Franklin was written in 1788, before the outbreak of the Revolution, and told of little but his longing for Paris and his memories of the happy days that he had spent there. Both of Black's letters, however, referred, among scientific matters, to "de Terray", and one referred to father and son of that name, who had visited Edinburgh, the father to bring his son to complete his studies. The older de Terray was the nephew of the famous Abbé Terray, former Minister of Finance and great-uncle of Mme. Lavoisier. The Terray family had held great place before the Revolution. Black also wrote of de Boulogne and de la Boulaye, who had arrived in Edinburgh; while we cannot identify these, one bears the name of a member of the Tax-Farm, Tavernier de Boulogne, a Farmer-General executed with Lavoisier. In the eyes of suspicion, are these not aristocrats and *émigrés,* whose safe arrival and comfort in Edinburgh are being reported? One of the letters indicates also that Lavoisier hoped to reach there himself.

The letter from Robert Kerr, English translator of Lavoisier's *Traité,* is harmless enough, but it refers to Sir James Hall, Baronet, and foreign aristocrats were no less hated than the native nobility by the faction in power in France. The letter mentioned also the bookseller Joseph de Boffe, of 7 Gerrard Street, Soho, London, possibly another *émigré.* The letter from Gillan to Mme. Lavoisier refers, among many other matters, to a visit to Paris of two Scots gentlemen of property, both received in "the first societies in London" and both wealthy, one the nephew of the former Governor of Madras. Visitors of this social class were suspect and so were all those who received them. The letter also made mention of Sir James and Lady Hall and Lord Daer. These names would not disarm suspicion, but suggest and even foster it. The letters from Méchain were, again, harmless enough, but they came from a member of the Academy of Sciences working inside the frontier of Spain, whose troops had now invaded France. It must have been very suspicious to the prejudiced eyes of those who read these documents.

Above all, however, the unsigned letter to Mme. Lavoisier, with date but no place of origin, was probably the most suspicious of the whole correspondence. It was dated September 11, 1792, and is apparently from a lady, a friend of Mme. Lavoisier, and was written from somewhere on the French side of the Spanish frontier. Its content was mainly political and there is no concealment of the writer's opinions on the current tyranny. She complains that her letters are opened and she sees herself and her friends menaced by horror and injustice. She has no news from Paris and is surrounded by "200 to 250 water-drinkers," presumably at some spa. Many rumors reach her, that the King is at Metz, and so on, but she believes none of them. She is horrified at the behavior of the *sans-culottes* at Orléans, and alarmed for the safety of someone whose identity she conceals. No one, she writes, will escape the barbarity and cruelty that must follow. She is anxious for her husband and her two sons. Perhaps she will go to Tarbes or Bayonne, and yet she has not lost hope of spending the winter in her own house. She urges Mme. Lavoisier to be courageous and a "brave Citizeness," ironically using the Revolutionary title. "This most beautiful revolution," she writes, "will make our streams run with blood and plunge us into total anarchy." But there are some comforts; for, even as she writes, the Prussians may be in Paris. What, one wonders, did the Committee of General Security make of that sentence? There was no real need to send that document to the Committee of Public Instruction for examination. The receipt, or even the mere possession, of this letter would have placed any Frenchman of that time, Farmer-General or not, in a position of the greatest peril from those who ruled his country.

If this letter, apart from the many references in the others, had been produced in evidence, it would have convicted of *incivisme* all concerned in it; and the references in the others would have "proved" correspondence about *émigrés,* counter-revolutionary activity, and complicity with the foreigner. As for the whole correspondence, it is difficult to believe that it could have been regarded as entirely innocent. Its various references to foreign aristocrats and to visits to Paris, the news of *émigrés* who had reached Scotland, Lavoisier's expressed hope of going there himself, and, above all, the sentiments of the unknown lady who wrote to Mme. Lavoisier, were all, slight though almost all of them are, items of evidence of antagonism to the Revolutionary Government. It is not improbable that these letters provided the concealed evidence that brought Lavoisier to his doom. Although we do not know what documents were removed from the houses of his colleagues during the search in September, 1793, we may well separate his case from theirs and perhaps find here the reason, admittedly hidden, why his old associates now in authority, especially Fourcroy and de Morveau, who had both been concerned in the official examination of these letters, the former on two occasions, made no attempts to save him from the scaffold.

Such an explanation is unhappily less incredible now than it would once have been. Why then, it may be asked, were the letters not produced at the trial? The trial was a farce; the evidence mattered little: and the real objects were the liquidation of doubtful elements, the acquisition of the fortunes of the Farmers-General, and the satisfaction of the public clamor raised against them for that purpose.

That foreign correspondence brings suspicion in such circumstances as then prevailed in France is undoubted. In the maddest moments and passions of that dismal and bloody hour, when the ruling faction mercilessly executed all whom they regarded as dangerous to their survival, all whom they wished to get out of the way, it is possible that the contents of some of these letters taken from Lavoisier's house may have provided the barely legal minimum of suspicion and formality that even the most desperate rulers seek in order ultimately to justify their actions. It is significant that these letters were not returned to Lavoisier. They remained in the possession of the Committee of Public Instruction, where they were twice examined by representatives. We have no record of any reports made as a result of these examinations.

Mme. Lavoisier, bereft at one stroke of husband, father, and many friends, and despoiled of her fortune, had little peace in her house in the Boulevard de la Madeleine. The confiscation of her husband's property by the State brought many visits from the officials of commissions and committees, on which he had been engaged, to recover the property belonging to these groups. Meanwhile, following the sentence of confiscation, official inventories had been ordered. On June 13 Quinquet had complained to the Temporary Commission of Arts that when he had gone to Lavoisier's house to obtain from the chemical laboratory various items that had been allocated for his official appointment, he had been hindered because the inventory had not been prepared. The Commission ordered immediate action. The next day Nicholas Leblanc (1742-1806), the chemist, and Berthollet began to prepare an inventory of the chemical apparatus and contents. But on the following day Mme. Lavoisier was arrested and detained, by order of the Committee of General Security, in the house taken for detainees in the Rue Neuve des Capucins. We do not know whether her arrest was in any way due to the letter that we have mentioned, and no reason is given by Grimaux. She was fortunate in obtaining from the local section of the Revolutionary Committee a favorable certificate for the Committee of Public Safety and the Committee of General Security, by which she obtained her freedom on August 17. On June 19, however, Leblanc and Berthollet continued their work at the house, but the task seems to have been difficult because of the necessity of removing the apparatus from cupboards and shelves where it had been stored and arranged in inadequate space.

A new report on September 24 by Dupin and the revising commissioners stated: that the Farmers-General owed the nation 130,345,262 livres, 12 sols, and 1 denier; that the nation had received only 67,360,090 livres, 21 sols, and 1 denier; that the property of their widows and their heirs must be used to pay this deficit; and that such of their children as had received dowries since 1775 must return them. As a result of the action that followed this report, Mme. Lavoisier lost her remaining income of 2,000 livres; she was, however, maintained by the work of an old family servant, Masselot.

Meanwhile, an old friend of the Lavoisiers—Pierre Samuel du Pont de Nemours (1739-1814), physician, Liberal economist—had been imprisoned. During these last anxious months du Pont had been in close touch with Lavoisier and had unavailingly urged him to leave Paris for Fréchines; he himself had for a time been forced to take refuge on his estate at Bois-des-Fossés, near Nemours. Financial difficulties, as well as political troubles, harassed him.

Du Pont's wife had died in 1782; he had two sons, the elder Victor, and the younger Eleuthère Irénée (1771-1834). Irénée had entered the Arsenal in 1787 to work with Lavoisier, leaving there on Lavoisier's resignation in 1792. With the coming of the Revolution, Pierre du Pont had lost his place and, in a Paris that now had freedom of the press and indulged in endless publication and pamphleteering, he turned to publishing. He was appointed publisher to the Academy of Sciences; the last annual volumes of the Academy's *Mémoires* carried his name and that of his son as printers. He intended to print the volumes of chemical memoirs that Lavoisier was compiling before and during his arrest. But du Pont had begun this venture without adequate resources, and had borrowed 710,000 livres from Lavoisier by mortgage on his house at Bois-des-Fossés at an interest of 4 per cent, the amount to be repaid in twelve years.

Early in August, 1794, three months after Lavoisier's execution, Pierre du Pont was arrested and taken to the prison of La Force. He expected from the Revolutionary Tribunal the same fate as that of his friend, but Robespierre had been guillotined on July 28 and reaction had set in. Du Pont was released in September. Immediately before and during his imprisonment, during part of which time Mme. Lavoisier also was in prison, his correspondence, as Professor S. J. French has shown, reveals his admiration of Mme. Lavoisier and a deep affection for her. A coolness had arisen between Mme. Lavoisier and the du Ponts, however, possibly because of the debt that could not be discharged owing to the difficulties that the troubled times had brought to publishing. But Mme. Lavoisier was now, in the autumn of 1794, without resources, having lost her remaining income by the decree of September 24. Professor French, in pointing out that repayment of any part of this debt would have meant its immediate confiscation to the State, expresses the view, after a study of the correspondence, that Pierre du Pont had a sincere affection for Mme. Lavoisier and was not attempting to cancel a debt by marriage, since he dare not pay that debt, even if he had been able to. A year later, Mme. Lavoisier having refused him, he married Mme. Poivre, widow of an old friend.

After the rise of Napoleon, du Pont, now without resources, resolved to emigrate. With his two sons and their families, he left in 1800 for America, where Irénée, finding that there was a demand for gunpowder in quantity and of good quality, started the first gunpowder factory in America on Brandywine Creek in Delaware. This was the initial step in the formation of the now world famous chemical combine of E. I. du Pont de Nemours and in the rise of America's great chemical industry—of which Lavoisier may therefore be regarded as the distant founder.

To return, however, to September, 1794, at 243 Boulevard de la Madeleine. The inventories now began in earnest. On September 27, the list of maps and plans was compiled by visiting officials appointed for the work. It included many well-worn sheets of Cassini's map of France, in boxes, each sheet in hinged sections. The inventory noted that Lavoisier had inserted symbols to indicate the minerals located at various places. There were also some sheets of Capitaine's well-known reduction of Cassini's map, and these were similarly marked. Altogether, the list comprised twenty-five cartographic items, many multiple, together with a model of a fully rigged ship. The maps were sent to the national repository of maps and plans.

Leblanc returned on November 9 to prepare a detailed inventory, since his earlier attempt on June 14 and 19 had not been satisfactory. He continued his work on November 11, 12, 17, 19, and 21. He was not accompanied by Berthollet, who may well have found the duty too distasteful and may have succeeded in evading it. It is known that Berthollet was afterwards welcomed at Mme. Lavoisier's receptions and that, apparently, she did not include him among those whom she blamed for their indifference to her husband's fate. The completed inventory of chemical apparatus and specimens showed over 13,000 items. A dealer in chemical apparatus and an apothecary had been ordered to accompany Leblanc to value the collection and they estimated it to be worth more than 7,000 livres.

While this work was in progress, on November 10, J. A. C. Charles (1746-1823), physicist and inventor of the hydrogen balloon, and Nicolas Fortin (1750-1830), maker of scientific instruments, together with Lenoir, arrived to prepare an inventory of the physical instruments and apparatus. They listed 125 items, many multiple, the actual total being about 250. An estimate of their value was not, it seems, completed, only ten pieces being valued and these at a total of 7,300 livres. The ten pieces included the three precision balances, one made by Fortin and two by Megnié; their value was set down as 3,500 livres, an estimate which was almost certainly made by Fortin and which probably indicates the price paid by Lavoisier. These were the historic instruments used in the determination of the standard weight in the new system of weights and measures; they were not actually in the house at this time, being deposited with the Commission of Weights and Measures.

From Leblanc's inventory, it is interesting to observe that Lavoisier had about 170 pounds of mercury, in 9 bottles, and about 60 pounds of mercuric oxide in 12 or 14 vessels, the whole being valued at 2,030 livres. The mercury was divided between the École Centrale (later the École Polytechnique), which received 5 bottles, and the Museum of Natural History in the Jardin du Roi (now the Jardin des Plantes), which received 4 bottles; each institution received also, 1 bottle of mercuric oxide.

The whole contents of the chemical laboratory were put at the disposal of the École Centrale, the Museum of Natural History, and the Bureau of Mines. The books in Lavoisier's library were also listed and many were allocated to different institutions; an inventory was also made of his collection of minerals. At Fréchines the same processes were in operation: crops were seized, furniture was sold, and books were confiscated. Items of furniture removed from 243 Boulevard de la Madeleine for the use of the Committee of Public Safety included 5 bookcases, a desk, a roll-top desk, and 6 chairs. Other furniture was taken for the Committee of Public Instruction. A piano, made by Zimmerman of Paris in 1786 and valued at 400 livres, was sent to the national repository of musical instruments on October 3, 1794; while on June 10 two carriages, a berline and a vis-à-vis, had been seized.

But, on December 10, 1794, the widows and children of the Farmers-General successfully petitioned the Convention to suspend further action until a report was presented to them, as they were reduced to poverty. When a representative, Lecointre, objected two days later that to revise one judgment would be to revise all, his argument was so vigorously disputed by Morellet in a pamphlet that on March 3, 1795, the Convention decided that the confiscated personal estate of the Farmers-General should be restored to their heirs and successors and that the sums realized by the items that had been sold should be passed to them. This decree did not, however, permit the return of their real estate. A representative in the Convention proposed an additional decree declaring that the order for the confiscation of this part of the property of the "unjustly condemned" Farmers-General should be annulled, and that the sequestration should be changed into a mere placing of seals until the accounts had been audited. The representative was Dupin, *Dupin-Mouillade,* who had so relentlessly pursued the Farmers-General to their doom, and one year earlier, to the very day, had denounced the extortions and adulterations of these "blood-suckers of the people."

But much had happened since then. On July 27, 1794, Robespierre had fallen, and the next day he was executed. The tide had turned and the Terror was over. Fouquier-Tinville had been executed after trial at the bar of the Tribunal where he had presided. While in prison awaiting execution, Vilate, one of his own jurymen, had produced an attack on him. Vilate had attacked Dupin also; he described him as "the Robespierre of the Farmers-General, the executioner of the tax-collectors, one of the tools of the tyrant," and reported that Dupin had said after the execution of his victims: "The guillotine is a better financier than Cambon." (Cambon was the printer of *assignats* and father of paper money.) Vilate, a spy in Robespierre's pay, had been denounced by Dupin and now denounced his denouncer in turn. The thieves had fallen out.

Dupin felt himself menaced, not merely by Vilate's pamphlet, but by the attention that Fouquier's trial had redirected to the trial of the Farmers-General. Thus he devised his bold plan to forestall his accusers. In the proposal that he now made to the Convention, he represented the trial of the Farmers-General as he wished it to appear, throwing all the odium on "the Robespierre faction, who had decided to coin money on the Place de la Révolution." He concluded: "I am more heart-broken than I can say in telling you that the decree that the Convention passed as a result of my report was the death-knell of the Farmers-General.—They were sent to their death without trial."

Attention was, however, concentrated on, rather than deflected from, Dupin after his intervention in the Convention. Mme. Lavoisier seems to have inspired the cry of vengeance which appeared against Dupin in a pamphlet, *Information by the Widows and Children of the former Farmers-General against the Representative-of-the-People Dupin;* corrected proofs of this document were found among her papers. It was an effective cry. The *Moniteur* of August 7 demanded that the Convention open an inquiry.

Dupin hurriedly announced that he would reply by positive facts to this fabric of suppositions and calumnies. But, as Grimaux pointed out, such a reply was impossible, because he had refused to consider the answers of the Farmers-General to the revising commissioners, because he had presented the statements made by these commissioners as proved facts, because he had deceived the Convention by concealing the defense presented by the Farmers-General, because he had demanded that they be arraigned before the Revolutionary Tribunal when he had no evidence of any counter-revolutionary activity on their part, and because he had communicated his report to Fouquier-Tinville and, as early as May 7, the day before the trial, he had demanded the preparation of an inventory of the contents of the offices of the Farmers-General.

When Dupin's reply was at last published, it was merely vague words. He tried to show that he had never supported Robespierre, he boasted of his humanity—the humanity, says Grimaux, of one who had pursued three forgotten Farmers-General to their death and who had deceived Didelot in advising him to give himself up at the Conciergerie; and Dupin added that he was so little concerned in the death of the Farmers-General that he could have brought fourteen others, whose names and whereabouts were known to him, to trial. Mme. Lavoisier's *Information* was presented to the Convention and a proposal was made that Dupin be arrested, not because of his false report, but as an assassin and a thief, for having denounced the three forgotten Farmers-General and for having stolen 100,000 livres in *assignats* and 95 gold louis from the wallet of Farmer-General de Lépinay. He was ordered into arrest and and deprived of his seat in the Convention on July 10, 1795. He strongly denied the theft alleged against him, which Grimaux regarded as far from proved. Set at liberty by the amnesty of October 26, 1795, he earned a living as a minor official of the revenue and died in obscurity in 1833.

Mme. Lavoisier obtained restitution of her husband's furniture, papers, books, scientific instruments, and laboratory apparatus, in the summer of 1795, the documents ordering restitution being inscribed "widow of the unjustly condemned Lavoisier." His books were returned —560 from the Bureau of Mines on July 13, and 134 from the Committee of Public Instruction, together with 41 from the local repository, on August 29. On July 27, the Bureau of Mines returned 3 hydrometers in their cases and on August 1, having failed to remove all the apparatus that had been allotted as their share, returned the 283 pieces that they had acquired. The physical instruments and apparatus were restored on August 21. All was carried out in order. The inventories had been so carefully compiled that Mme. Lavoisier was able, for example, to reconstitute the library at Fréchines with only three volumes missing.

The liquidation of the Tax-Farm was entrusted to a new audit commission and went on for several months. Sequestration of the property of the Farmers-General was simplified into mortgage, which was finally cancelled in 1806. A decree of the Council of State established that the Farmers-General, far from owing the nation 130 million livres, were the Nation's creditors to an amount of 8 million livres, the final answer to all the charges brought against them, but none of their heirs laid claim to any part of this sum.

For some time Mme. Lavoisier was engaged in the preparation of her husband's uncompleted memoirs for publication. He had started this work in 1792 and printing had begun in April, 1793; with Séguin's help he was correcting the first proofs during his detention. He had originally intended to publish eight volumes, which would include his memoirs, together with those of other scientists whose work had helped to establish the new chemistry. When he died, only two volumes were ready and part of a third. These fragments were published privately in two volumes in 1805, Mme. Lavoisier justifiably refusing to accept a pretentious introduction from Séguin, who claimed for himself a great share in the work, and Séguin declining to compose a preface branding those responsible for Lavoisier's death. Copies of these two volumes were presented to many eminent scientists in Europe and to the great scientific societies. None were sold, except those left at her death.

In 1805, at the time of the publication of these memoirs, Mme. Lavoisier had left her old house in the Boulevard de la Madeleine and was living in the Rue d'Anjou, also in the Faubourg St.-Honoré. The house to which she had moved was demolished in 1841 and the Rue Lavoisier now runs where it formerly stood. Here she began once more to hold her receptions. Her salon was frequented by the leaders of science in France—Delambre, Cuvier, Prony, Lagrange, Laplace, Berthollet, Arago, Biot, Humboldt, and others, but never by those who had failed to use their political influence to save her husband from the guillotine.

Among her guests was Sir Charles Blagden (1748-1820), whom we have already encountered as a visitor to Lavoisier's laboratory at the Arsenal and who was the medium by whom Lavoisier was informed in 1783 of Cavendish's experiments on the burning of inflammable air and the resultant formation of water. Blagden was elected Fellow of the Royal Society in 1772 and became secretary of the Society in 1784. He was a physician and a medical officer in the Army until 1814, but he spent much of his time in Paris, enjoying its social and cultural life. He proposed marriage to Mme. Lavoisier, but was rejected.

Another guest was Sir Benjamin Thompson, Count von Rumford, born in North Woburn, Massachusetts, in 1753, student of John Winthrop in Harvard College, and schoolmaster in Wilmington, Delaware. In 1775, after being suspected of opposition to the national aspirations for independence and after arrest and imprisonment, he left for England and took some part on the British side in the War of American Independence. In England he became Under-Secretary for the Colonies in 1780 and he was knighted in 1784. During the years from 1784 to 1795 he rendered such valuable service to the Elector of Bavaria in leading that state out of its economic difficulties that he was created Count von Rumford. He had made many outstanding contributions to science, especially on the subject of heat; and he was a prominent Fellow of the Royal Society, to which he was elected in 1779. After the death of the Elector, Rumford returned to London, where in 1799 he founded the Royal Institution, first of all research institutes, but in 1803 he left England for France. His interests closely resembled those of Lavoisier, both in science and its applications and also in social and economic problems. On October 22, 1805, Mme. Lavoisier was married to Count von Rumford, insisting, however, on being known as the Countess Lavoisier-Rumford. The marriage proved unhappy and after four years of mutual discontent there was an agreement to separate. Rumford died in 1814.

Mme. Lavoisier lived on for many years, her salon frequented as before, but she died suddenly on February 10, 1836, at the age of seventy-eight, forty-two years after her husband's execution. Grimaux failed to trace any descendant of the original Lavoisier family of Villers-Cotterets; and so, with the death of Mme. Lavoisier, the last personal link with the founder of modern chemistry was severed.

Unlike his great predecessor Newton, Lavoisier had witnessed in his own lifetime the acceptance of the new science that his genius had created. In the summer of 1943 in the Palais de la Découverte in Paris, not far from the scene of the tragedy of May 8, 1794, an exhibition of Lavoisier's scientific apparatus, papers, and personal relics, was held to mark the two-hundredth anniversary of his birth in a celebration that breathed the spirit of France and the glory of the French intellect: already under the heel of the invader for three years and still anxiously awaiting her deliverance, France remembered with pride the enduring triumphs of one of her greatest sons.

BIBLIOGRAPHY

Sources for Lavoisier's life and works:

BERTHELOT, M. *La Révolution chimique–Lavoisier.* Paris, 1890. A study of Lavoisier's laboratory notebooks, with extensive quotations.

GRIMAUX, E. *Lavoisier 1743–1794 d'après sa correspondance, ses manuscrits, ses papiers de famille et d'autres documents inédits.* Paris, 1888; 2nd edn., Paris, 1896; 3rd edn., Paris, 1899. The standard life of Lavoisier, but only about thirty pages are allotted to his scientific work; and, unfortunately, although it is documented, it contains many errors or misprints, especially of figures such as dates and references.

LAVOISIER, A. L. *Opuscules physiques et chymiques.* Paris, 1774. English translation by Thomas Henry. *Essays Physical and Chemical.* London, 1776.

——— *Traité élémentaire de Chimie.* Paris, 1789. English translation by Robert Kerr. *Elements of Chemistry.* Edinburgh, 1790.

——— *Œuvres de Lavoisier publiées par les soins de son Excellence le Ministre de l'Instruction Publique et des Cultes.* Paris, 1862-1893, 6 vols. Vol. I, ed. Dumas, 1864, contains the *Opuscules* and the *Traité;* vol. II, ed. Dumas, 1862, includes the memoirs on chemistry and physics; vols. III and IV, ed. Dumas, 1865 and 1867, include various other memoirs and reports; vol. V, ed. Grimaux, 1892, contains studies on geology and mineralogy, etc., and memoirs and reports relating to the Commission for Gunpowder; and vol. VI, ed. Grimaux, 1893, contains other reports, together with studies on agriculture, politics, and economics, and documents relating to the Tax-Farm and the Commission of Weights and Measures.

Méthode de Nomenclature chimique proposée par MM. de Morveau, Lavoisier, Bertholet [*sic*], *& de Fourcroy.* Paris, 1787. English translation by James St. John. *Method of Chemical Nomenclature.* London, 1788.

Mémoires de l'Académie Royale des Sciences. Paris. Volumes from 1765 to 1790.

Observations sur la Physique. Paris. Ed. Rozier. Volumes from 1771 to 1783

Literature preceding and contemporary with Lavoisier:

BLACK, JOSEPH. *Dissertatio medica inauguralis, de Humore Acido a Cibis orto, et Magnesia Alba.* Edinburgh, 1754.

——— *Experiments upon Magnesia Alba, Quicklime, and some other Alcaline Substances.* Edinburgh, 1756.

BOYLE, The Hon. ROBERT. *Essays of Effluviums.* London, 1673.

——— *New Experiments Physico-Mechanicall* etc. Oxford, 1660.

——— *The Sceptical Chymist.* London, 1661.

CAVENDISH, HENRY. Experiments on Air. *Philosophical Transactions,* 1784, 74: 119.

——— Experiments on Factitious Air. *Philosophical Transactions,* 1766, 56: 141.

HALES, STEPHEN. *Vegetable Staticks.* London, 1727.

VAN HELMONT, JOHANN BAPTISTA. *Ortus Medicinæ.* Amsterdam, 1648. English translation by J [ohn] C [handler]. *Oriatrike.* London, 1662.

HOOKE, ROBERT. *Micrographia.* London, 1665.

MAYOW, JOHN. *Tractatus quinque medico-physici.* Oxford, 1674.

PRIESTLEY, JOSEPH. *Experiments and Observations on Different Kinds of Air.* London, 1774-75-77, 3 vols.

——— Letter to Sir John Pringle, P.R.S. *Philosophical Transactions,* 1775, 65: 384.

REY, JEAN. *Essays de Iean Rey* etc. Bazas, 1630. Facsimile reprint with introduction by D. McKie. London, 1951.

SCHEELE, CARL WILHELM. *Chemische Abhandlung von der Luft und dem Feuer.* Upsala und Leipzig, 1777. English translation by Leonard Dobbin in *The Collected Papers of Carl Wilhelm Scheele.* London, 1931.

Biographies:

DAUMAS, MAURICE. *Lavoisier.* Paris, 1941.

FRENCH, S. J. *Torch & Crucible. The Life and Death of Antoine Lavoisier.* Princeton and London, 1931.

GRIMAUX, E. *Lavoisier 1743-1794* etc., as above.

HOLT, ANNE. *A Life of Joseph Priestley.* London, 1931.

MCKIE, D. *Antoine Lavoisier, the Father of Modern Chemistry.* London, 1935. An analysis of Lavoisier's contribution to chemistry.

THORPE, T. E. *Joseph Priestley.* London, 1906.

WILSON, G. *The Life of the Hon.ble. Henry Cavendish.* London, 1851.

Other works, special and general, and recent studies:

BERTHELOT, M. *Notice historique sur Lavoisier.* Paris, 1889.

ANTOINE LAVOISIER

Hartog, Sir Philip. The Newer Views of Priestley and Lavoisier. *Annals of Science,* 1941, 5: 1-56.

Lenglen, M. *Lavoisier agronome.* Paris, 1936.

Meldrum, A. N. *The Eighteenth-Century Revolution in Science. The First Phase.* Calcutta, 1929.

——— Lavoisier's Early Work in Science. *Isis,* 1933, 19: 330-363; 1934, 20: 398-425.

McKie, D. and Heathcote, N. H. de V. *The Discovery of Specific and Latent Heats.* London, 1935.

Partington, J. R. *The Composition of Water.* London, 1928.

——— *A Short History of Chemistry.* London, 1937; 2nd edn., London, 1948.

Partington, J. R. and McKie, D. Historical Studies on the Phlogiston Theory. *Annals of Science,* 1937, 2: 361-404; 1938, 3: 1-58 and 337-371; 1939, 4: 113-149.

White, J. H. *The History of the Phlogiston Theory.* London, 1932.

Wolf, A. *A History of Science, Technology and Philosophy in the Eighteenth Century.* London, 1938; 2nd edn., London, 1952.

De la Rivière, R. Dujarric. *Lavoisier Économiste.* Paris, 1949.

INDEX

A

INDEX

P

Q

R

S

T

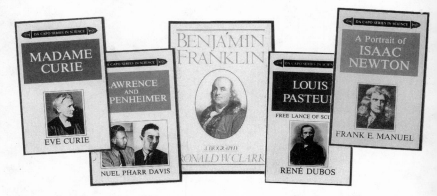